Advancements in Forensic DNA Analysis

Hirak Ranjan Dash • Kelly M. Elkins •
Noora Rashid Al-Snan

Advancements in Forensic DNA Analysis

Springer

Hirak Ranjan Dash
Forensic Biology and Biotechnology
National Forensic Sciences University
Delhi, Delhi, India

Kelly M. Elkins
Department of Chemistry
Towson University
Towson, MD, USA

Noora Rashid Al-Snan
DNA Forensic Laboratory
Ministry of Interior
Manama, Bahrain

ISBN 978-981-99-6194-8 ISBN 978-981-99-6195-5 (eBook)
https://doi.org/10.1007/978-981-99-6195-5

This Springer imprint is published by the registered company Springer Nature Singapore Pte Ltd.
The registered company address is: 152 Beach Road, #21-01/04 Gateway East, Singapore 189721, Singapore

Preface

DNA fingerprinting technique is considered as the most irrefutable evidence in the criminal justice system. Since its inception, the technology is evolving at a rapid pace. In the last few decades, the technology has been advanced to fast protocols, increase in the discrimination power, sensitivity, and the ability to analyze challenging samples. With the increase in the number of STR markers to be analyzed, new commercial kits are being launched. DNA extraction and quantification protocols have been revolutionized, resulting in the ability to analyze minute quantities of DNA for further examination. Further, developments in the field of criminal and crime scene database and probabilistic genotype software as well as DNA mixture interpretation have revolutionized forensic DNA analysis in identifying suspects from crime scene samples. The technology saw a paradigm shift with the use of Next Generation Sequencing (NGS) technology in forensic DNA analysis. The NGS technology promises age estimation, body fluid identification, forensic genealogy, DNA phenotyping, detection of geographic origin, and ancestry of an individual and many others. As the need for a technological shift has become imperative, students and working professionals should be aware of these advanced technologies. This book aims to provide a handful of knowledge to DNA practitioners, researchers, students, police and law professionals as well as students.

The book is written in simple English that requires a basic background in biological science to understand. Hence, it will be helpful for student communities of biology, zoology, wildlife, medicine, anthropology, microbiology, forensic science, genetics, and law fraternity at the graduate, postgraduate, upper-level undergraduate, and research level. This book will be a handy reference among scientific communities to quickly explain and summarize technological advancements in the field of forensic DNA analysis. Investigating agencies and the law practicing partners can gather a sound knowledge on the use of various innovative technologies in the criminal justice system.

Being in different geographical time zones, it was not an easy job for us to bring out this book. We have spent many sleepless nights to gather for meetings and quick discussions while preparing the book. We would like to take this opportunity to acknowledge the family members, friends, and relatives who have shown unconditional support during the preparation of the book. We hope this book will meet the expectations of the reader.

Wishing a happy reading!!!

Delhi, India — Hirak Ranjan Dash
Towson, MD, USA — Kelly M. Elkins
Manama, Bahrain — Noora Rashid Al-Snan

Contents

About the Editors

Hirak Ranjan Dash is an Assistant Professor of Forensic Biology and Biotechnology at National Forensic Sciences University, India. Before joining academics, he has worked as a Forensic DNA expert at Forensic Science Laboratory, Madhya Pradesh, India. He has examined more than 1000 cases using the DNA fingerprinting technique. His research interests include DNA fingerprinting, microbial forensics, microbiome analysis, next generation sequencing, and mtDNA analysis. He has published more than 40 research papers in various peer-reviewed journals. He has written 10 books in various fields of biotechnology. He is a life member and served as a reviewer for various international journals. He is the pioneer worker from India on NGS-based forensic DNA analysis. He is featured in the list of top 2% scientists of the world for 2021, 2022, and 2023.

Kelly M. Elkins is a Professor of Chemistry at Towson University, USA, founding co-editor-in-chief of an international journal, and former Director of the Forensic Science Program at the Metropolitan State College of Denver. She is a Fellow of both the American Academy of Forensic Sciences and American Chemical Society. Her research interests include improving DNA recovery, applying forensic DNA typing to historic human remains, development of assays for bioterror agents, detection and identification of novel psychoactive substances and plant drugs, and ethics in forensic science. She has published more than 60 book chapters and research papers in peer-reviewed journals and 7 books on various topics in forensic chemistry and forensic biology. She has consulted on paternity, unidentified human remains, and other unsolved cases. Her work has been funded by numerous agencies and she was featured in the list of top 2% scientists of the world for 2019.

Noora R. Al-Snan is an enthusiastic DNA analyst who has joined the Ministry of Interior (MOI) in the General Directorate of Criminal Investigation and Forensic Science, FSL (Bahrain), in 2008. Currently, she is assigned as the head of Biology & DNA Forensic Lab. In 2019, she has completed her PhD degree in molecular medicine specializing in medical genetics in the topic of DNA typing for the Bahraini population under the supervision of the Arabian Gulf University (AGU) with the cooperation of Naif Arab University for Security Sciences (NAUSS) (Riyadh, KSA). In 2012, she has received her

MSc degree in medical biotechnology specializing in genetic engineering for recombinant proteins from AGU. She has completed a bachelor's degree in molecular biology and biochemistry studies from Kuwait University (Kuwait). In her 15 years of forensic experience, she has analyzed and solved many challenging cases mainly related to terrorism, robbery, and murder. Dr. Noora has attended many conferences and workshops as a speaker and delegated from Ministry of Interior (MOI). She has achieved many outstanding results along with her team, including obtaining the first position worldwide for DNA hit of the year (2023) in bomb cases and previously (2017) awarded the fifth position worldwide from the same organization: The Gordon Thomas Honeywell DNA during the HIDS Conference. In addition, one of the published scientific papers was the most frequently downloaded paper in *Molecular Genetics and Genomics*. Also, she is corresponding author for several scientific papers and books and international reviewer for many published articles. Al-Snan is a member of the scientific committee in NAUSS as well as in ISFG and ISFG Arabic speaking group. She has set innovative protocols for the recovery of DNA from RDX-C4 evidence, Forensic DNA intelligence, and fast and effective DNA extraction from teeth, bones, and maggots and has innovative strategy to obtain investigative leads from homozygous loci count. Her main interests are investing in human resources, continuous education, and R&D.

Current Status and Advancements of Forensic DNA Analysis

1

1.1 Introduction to Forensic DNA Analysis

DNA is widely used as a tool in criminal and civil legal cases. Forensic cases include sexual assaults, murders, suicides, mass disasters, terrorism events, document-tampering, burglaries, and other cases of theft. Since its application to forensic science 40 years ago, the collection and analysis of DNA has become nearly ubiquitous in legal cases. Juries expect DNA evidence and analysis. Lawyers associate DNA with reliability.

The application of DNA to forensic science emerged from genetics research and clinical applications (Patrick 2007). Some DNA sequences and repeating units were observed to vary significantly among individuals. Loci with variation among individuals were substantially researched and those loci found to have the lowest rates of change or mutation rates and the highest heterozygosity were selected as the targets for a human identification panel (Butler 2005). In the earliest days, single loci were investigated in parallel or sequentially in separate assays. Probing loci individually was slow and time-consuming. Scientists quickly focused on multiplexing the target sites (Butler 2005). Shortly thereafter, duplex, triplex, and quadruplex assays emerged. The multiplexing has continued although further improvements are contingent upon technological advances (Butler 2005). The addition of next generation sequencing (NGS) to the repertoire of techniques used in forensic science has greatly increased the loci that can be typed simultaneously and overcome many technical problems to further multiplexing by the now standard method of capillary electrophoresis (CE). Furthermore, NGS has been demonstrated to be more sensitive facilitating more complete DNA analysis for challenging samples such as ancient and historic human remains, human remains burned in arson and mass disaster cases, poorly stored or old evidence items, hairs, and items with only trace cellular material (Elkins et al. 2022). More complete DNA profiles is an important accomplishment toward solving more cases. Many of the advances have occurred in the past 5 years; these are the focus of this book.

It has been observed that individuals who commit a crime often commit additional crimes. In fact, approximately 40% of criminal offenders are repeat offenders. For this reason, law enforcement agencies, with support from national and international bodies, have created databases of DNA analysis results. Nowadays, law enforcement databases including the U.S. Combined DNA Index System (CODIS) include DNA typing results for DNA recovered from items in unsolved cases, criminal offenders, arrestees, detainees, unidentified human remains, as well as missing persons. The sensitivity of modern DNA techniques and the size of the databases is enabling law enforcement to solve decades-old cases.

Many excellent books have been written for the student or newcomer to forensic DNA

H. R. Dash et al., *Advancements in Forensic DNA Analysis*,
https://doi.org/10.1007/978-981-99-6195-5_1

analysis in the past (Butler 2005; Elkins 2013; Elkins and Zeller 2021). However, there have been numerous advancements and changes in the past 5 years. Herein, you will find chapters on the technological advances, applications of genetic techniques to the identification of body fluids and non-human DNA, the role of DNA databases in forensic investigations, probabilistic statistics, casework applications, and ethical questions regarding DNA analysis and especially the application of investigative genetic genealogy (IGG) in forensic casework. In case you are new to forensic DNA analysis, we shall review the DNA typing process. We will describe the current status of the techniques in each step of the process. We will describe the technological advancements that support each advancement in biological profiling. We will close with a final chapter on the future directions of DNA analysis in forensic science and ethics of forensic DNA typing (DeAngelo and Elkins 2023).

1.2 An Overview of DNA Typing Targets and Approaches

The DNA typing process has several steps. The focus of standard DNA typing has been short tandem repeats (STRs). STRs are short, repeated, and inherited elements in the genome. They have been found on the autosomal chromosomes as well as the X and Y sex-determining chromosomes. For human forensic DNA typing for individualization, labs do not sequence the complete 3.2 billion base pair genome; rather the focus has been on counting tetranucleotide repeats, although there are di-, tri-, and penta-nucleotide repeats and longer repeats in the genome as well (Edwards et al. 1991).

Tetranucleotide repeats were chosen because the repeats are relatively short so they can be easily copied by the polymerase chain reaction (PCR). The are many polymorphic STR loci and the four-base pair repeat is relatively easy to size on gels and using CE. Loci were chosen from across the chromosomes that make up the human genome so that unlinked loci are evaluated for downstream statistical analysis. Sanger sequencing can be used to directly type the STR but sizing is more widely used because it is faster and multiple loci can be investigated simultaneously by multiplexing. When DNA is sized using CE, an internal standard is added to the evidence DNA sample so that parallel CE runs can be directly compared. The sample is amplified by PCR using a set of primers that copy the DNA at one or several chosen STR loci. One of the primers for each locus is tagged with a fluorescence dye so that the amplicon contains the tag. Following PCR, the amplicons are separated using CE with a relatively short (47 cm) capillary and POP-4 polymer (Butler 2005). A DNA ladder containing all of the possible alleles for each of the loci is separated in a separate capillary. The resulting data is an electropherogram. Each allele corresponds to a DNA of the length of a STR repeat number identified at the locus in human population studies. The number of repeats identified in these studies varies across loci. Using the internal standard added to both the sample DNA and the DNA ladder, the elution times can be directly compared and the number of repeats, or allele, can be assigned for each locus using the DNA ladder. If only a few cells were recovered from an evidence sample, the amplicon may be undetectable using this method (Butler 2005).

Other types of inherited elements have also been investigated for forensic DNA typing. While STRs vary by length and long stretches of DNA are typically copied and sized to determine the allelic profile; only one base must be identified to identify a single nucleotide polymorphism (SNP). SNP loci can also be sequenced individually by Sanger sequencing but faster approaches including SNaPshot assays, PCR high resolution melt analysis, and NGS are now more frequently used. SNaPshot assays have been shown to type ten SNPs simultaneously (Butler 2005) while hundreds and thousands of SNPs can be typed simultaneously using NGS (Jäger et al. 2017). Microhaplotype loci (microhaps, MHs) are a newer addition to the loci that can be typed for forensic use. These markers are short (<300 bp) composed of two or more SNPs in close proximity with multiple allelic types (Oldoni et al. 2019). These have been detected by NGS.

1.3 An Overview of the DNA Typing Process with Advancements

Figure 1.1 shows an overview of the DNA typing process with advancements including NGS and genetic genealogy. After evidence is collected at the crime scene, it is submitted for analysis to the crime laboratory. The evidence is visually screened and may undergo serological or other testing including fingerprint analysis prior to being sampled and chosen for DNA typing. In the next steps of the process, the DNA is extracted from an evidence sample, quantitated, and typed using the standard CE method. Normally, after this step, the DNA profile is determined, uploaded to a local or state database and finally to a combined national database such as CODIS, and compared to existing profiles in the database. Statistics are applied and a random match probability, likelihood ratio, or probability of exclusion statistic is computed. Based on the statistical result, the sample is determined to have a probability of deriving from a source or being excluded (Butler 2005; Elkins 2013).

In many cases, however, a complete DNA profile is not obtained using the standard CE DNA typing approach. In these cases, samples may be selected for advanced forensic DNA analysis by NGS (Elkins and Zeller 2021). Other reasons a sample may be typed with NGS is that law enforcement is interested in developing a phenotypic profile of externally visible characteristics (EVCs) or predict the ancestry of the person who deposited the biological material on the evidence item. Or paternity testing or familial lineage testing is sought. NGS can also be used to characterize biological weapons in bioterrorism events (Minogue et al. 2019). Commercial NGS kits also include STRs located on the X- and Y-chromosomes. X- and Y-STR loci are included in a commercial NGS kit with standard autosomal STRs used for identity testing (Jäger et al. 2017) whereas for CE, they are generally in separate kits and must be tested separately from autosomal STRs. Furthermore, in the NGS DNA typing kits, SNP loci are targeted along with STR loci. The ForenSeq kit analyzes up to 58 STRs and 172 SNP loci. With CE, the STRs are not routinely sequenced; the profile is assignment based on the amplicon sizes. With NGS, STRs and SNPs and surrounding flanking regions are sequenced. NGS has been shown to be reveal more variations in the loci. Not only are STRs and SNPs typed but additional microhaplotypes and SNP variations have been uncovered in the STR loci and flanking regions. Furthermore, NGS is a direct technique and low level DNA is sequences and the sequences are output for analysis making it easier to uncover low template DNA contributors and mixed source samples.

The newest NGS kit for forensic applications types 10,230 SNPs simultaneously. This data can be used to solve cold cases. The data can be uploaded to a database called GEDmatch. GEDmatch contains DNA profiles collected with personal genomics ancestry kits from 23andMe, MyHeritage, AncestryDNA, LivingDNA, Findmypast, and FamilyTreeDNA and uploaded by users for additional genealogy searching using gene sequences. Stretches of sequence data are compared across chromosomes until matches are found. Individuals with more regions in common are likely to have closer ancestry. From the genetic screening, a common ancestor can be identified and modern descendant relatives of the person who deposited the evidence samples can be identified. Using additional information including known workplace, credit card, phone, and travel records, investigators can eliminate individuals that do not match the pattern or known data for the person in question until they make an identification or determine that there is no match. The U.S.-based database is populated with profiles primarily of Americans of European-descent at this time so cases involving people of color are less likely to be solved (Rodriguez et al. 2021).

Rapid DNA typing process and instruments have been developed for investigations and

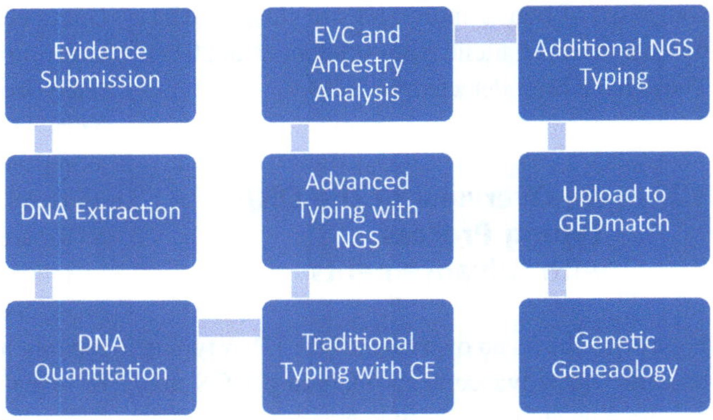

Fig. 1.1 An overview of the advanced DNA typing process

questions of parentage at international borders, in immigration cases, and for roadside investigations. The samples is collected at the scene, inserted into the instrument and a DNA profile is obtained in 90 min.

Most recently, the DNA typing process has been miniaturized on microfluidic devices (Turiello et al. 2023).

1.4 An Overview of the Advancements in Chemistry and Technology

Early detection approaches in forensic DNA typing included silver stain (Butler 2005) although they could be detected by a fluorescent intercalating dye. Fluorescence dyes were widely adopted as they vary in their emission maxima (Butler 2005; Elkins 2013). The number of dyes employed increased, thereby increasing the number of loci that can be multiplexed and simultaneously detected and analyzed (Table 1.1). The first locus was HLA-DQα in 1986 and the so-called monoplex commercial STR kit targeting TH01 manufactured by Promega was released in 1993. Several kits followed that were developed by multiple manufacturers including Promega, Applied Biosystems (now Thermo Fisher), and Qiagen. The kits were developed to analyze the loci adopted by the European Union in the European Standard Set (ESS), Interpol in the Interpol Standard Set of Loci (ISSOL), China, and the U.S. for CODIS. Seven loci were adopted by the ESS in 2001 and five additional loci were added in 2009. Thirteen CODIS loci were adopted in 1997. As technology improved to detect and deconvolute more dyes, more primer sets were multiplexed to reduce the number of assays that needed to be run to analyze all of the adopted STR loci simultaneously. Y- and X-STR kits were developed to probe loci on the sex-determining chromosomes for to aid in answering lineage questions. After the release of the Applied Biosystems 3500 CE instrument capable of detecting and analyzing six dyes, multiplexes were developed that utilized five dyes for the loci and one for an internal standard. In the U.S., seven additional STR loci were added to the standard DNA typing panel effective January 1, 2017 (Hares 2015; Scientific Working Group on DNA Analysis Methods 2017; Elkins and Zeller 2021). New 6-dye STR kits were created to meet the needs of the new requirements. These included the ESS and CODIS loci and a couple of sex-determining markers. Promega introduced the 24-plex Fusion in 2012 and the PowerPlex® Fusion 6C System developmental validation was published in 2016 (Ensenberger et al. 2016). Applied Biosystems introduced GlobalFiler in 2012. Qiagen introduced its Investigator 24plex GO! and Investigator 24plex DS kits in 2017 followed by its Investigator 26plex QS. The SureID PanGlobal kit was validated in 2017 (Liu et al. 2017). The QS kits have a quality sensor. Most recently, Promega released its PowerPlex® 35GY System with eight dyes in 2022 and Spectrum CE system to handle it.

The first commercial NGS kits for forensic applications were introduced in 2014 by Applied

Table 1.1 Advancements in multiplexing of STR loci for CE analysis using fluorescence dyes for forensic applications (https://strbase.nist.gov/multiplx.htm)

Year	STR kit	Manufacturer	Number of loci	Number of dyes
1993	TH01	Promega	1	1 or silver stain
1994	CTT	Promega	3	1 or silver stain
	CTTv	Promega	4	1 or silver stain
1994	FFv	Promega	3	1 or silver stain
	FFFL	Promega	4	1 or silver stain
	SilverSTR™ III Multiplex	Promega	3	1 or silver stain
1996	GammaSTR™ Multiplex	Promega	4	1 or silver stain
1996	PowerPlex™ System	Promega	8	3
1996	AmpF*l*STR® Blue	Applied Biosystems	3	1
1997	AmpF*l*STR® Green I	Applied Biosystems	4	1
1997	AmpF*l*STR® Profiler	Applied Biosystems	10	3
1997	AmpF*l*STR® Profiler Plus®	Applied Biosystems	9	3
1997	PowerPlex™ 1.1	Promega	13	3
1998	PowerPlex™ 1.2	Promega	8	3
1998	AmpF*l*STR® COfiler®	Applied Biosystems	7	3
1999	PowerPlex™ 2.1	Promega	9	3
1999	AmpF*l*STR® SGM Plus®	Applied Biosystems	11	4
2000	PowerPlex™ 16	Promega	16	5
2001	AmpF*l*STR™ Identifiler™	Applied Biosystems	16	5
2001	AmpF*l*STR™ Profiler Plus® ID	Applied Biosystems	10	4
2002	AmpF*l*STR™ SEfiler	Applied Biosystems	12	5
2002	PowerPlex™ ES	Promega	9	4
2003	PowerPlex™ Y	Promega	12	3
2004	AmpF*l*STR™ Yfiler	Applied Biosystems	16	5
2007	AmpF*l*STR™ MiniFiler	Applied Biosystems	9	5
2007	AmpF*l*STR™ SEfiler Plus®	Applied Biosystems	12	5
2007	PowerPlex™ S5	Promega	5	3
2008	AmpF*l*STR™ Sinofiler	Applied Biosystems	16	5
2009	AmpF*l*STR™ Identifiler™ Direct	Applied Biosystems	16	5
2009	AmpF*l*STR™ NGM	Applied Biosystems	16	5
2009	PowerPlex™ 16 HS	Promega	16	5
2009	PowerPlex™ ESX 16	Promega	16	5
2009	PowerPlex™ ESX 17	Promega	17	5
2009	PowerPlex™ ESI 16	Promega	16	5
2009	PowerPlex™ ESI 17	Promega	17	5
2009	PowerPlex™ CS7	Promega	7	3
2010	AmpF*l*STR™ Identifiler™ Plus	Applied Biosystems	16	5
2010	AmpF*l*STR™ NGM SElect	Applied Biosystems	17	5
2018	AmpF*l*STR™ VeriFiler Plus	Applied Biosystems	27	5
2010	Investigator ESSplex Plus	Qiagen	16	5
2010	Investigator ESSplex SE	Qiagen	17	5
2010	Investigator ESSplex SE QS	Qiagen	17	5
2010	Investigator IDplex Plus!	Qiagen	16	5
2010	Investigator IDplex Go!	Qiagen	16	5
2010	Investigator Nonaplex ESS	Qiagen	9	4
2010	Investigator Hexaplex ESS	Qiagen	7	4
2010	Investigator HDplex	Qiagen	13	4
2010	Investigator Triplex AFS QS	Qiagen	3	4

(continued)

Table 1.1 (continued)

Year	STR kit	Manufacturer	Number of loci	Number of dyes
2010	Investigator Triplex DSF	Qiagen	3	3
2010	Investigator Argus X-12 QS	Qiagen	14	5
2010	Investigator Argus Y-12 QS	Qiagen	11	5
2011	PowerPlex™ 18D	Promega	18	5
2012	PowerPlex™ Y23	Promega	23	5
2012	PowerPlex™ 21	Promega	21	6
2012	GlobalFiler™	Applied Biosystems	24	6
2012	PowerPlex® Fusion	Promega	24	6
2017	Investigator 24plex GO!	Qiagen	24	6
2017	Investigator 24plex DS	Qiagen	24	6
2021	Investigator 26plex QS	Qiagen	26	6
2021	Investigator Argus Y-28 QS	Qiagen	27	6
2022	PowerPlex® 35GY System	Promega	32	8

Biosystems and several additional products have been introduced by Verogen and Qiagen. The NGS kits were developed to be backward compatible and simultaneously probe the ESS, Interpol and CODIS markers while adding several new functions. SNP loci were added to the multiplex STR panels for predicting ancestry and phenotype. The Amelogenin loci was retained and X- and Y-STR loci were added. Applied Biosystems introduced the HID-Ion AmpliSeq™ Ancestry Panel of 165 autosomal SNPs and the HID-Ion AmpliSeq™ Identity Panel of 124 SNPs in 2014 (Table 1.2).

In 2016, Promega introduced the PowerSeq 46GY kit covering 22 autosomal STRs, 23 Y-STRs and Amelogenin. In 2018, Thermo Fisher introduced the Precision ID kit covering 35 markers including 30 autosomal STR markers, Amelogenin, 1 Y-STR, 2 additional Y markers, and 1 SNP. In 2017, Verogen introduced the ForenSeq kit with two primer set options, DNA primer mix A and B (DPMA and DPMB). DPMA contains primer pairs for 58 STRs (including 27 autosomal STRs, 7 X-, and 24 Y- STR markers) and 94 identity informative (ii) SNPs for a total of 152 loci. DPMB contains primer pairs for all of the DPMA loci as well as 22 phenotype SNPs (pSNPs) and 56 ancestry informative (ai) SNPs (two aiSNPs are also used for phenotype estimation) for a total of 230 loci. In 2021, Verogen released the MainstAY and Amelogenin and Kintelligence kits. The MainstAY kit probes 53 loci including 27 autosomal STRs, 25 Y-STRs, and Amelogenin. The Kintelligence kit probes 10,230 SNPs. In 2022, Verogen released the Imagen kit with two primer set options, DPME and DPMF, for phenotype predictions. DPME includes 22 hair color SNPs, 6 eye color SNPs, 36 skin tone SNPs and 14 Y-SNPs for a total of 55 SNPs. DPMF includes 22 hair color SNPs, 6 eye color SNPs, 36 skin tone SNPs and 56 SNPs for biogeographical ancestry prediction as well as 14 Y-SNPs for a total of 111 SNPs. Some of the SNPs are new to the Verogen lines and not included in DPMB. NGS conducted on Ion Torrent and MiSeq instruments was approved for CODIS upload in 2019. Precision ID and ForenSeq™ Signature Prep kit data were approved for upload and inclusion in CODIS (Scientific Working Group on DNA Analysis Methods Addendum 2019). Qiagen has created several NGS products as well including the QIAseq Investigator ID SNP panel, QIAseq Investigator Global Ancestry SNP panel, QIAseq Investigator Middle East Ancestry SNP panel, QIAseq Investigator Microhaplotype panel, and the QIAseq Investigator Missing Persons SNP panel. The QIAseq Investigator ID SNP panel was reported to have 140 SNPs in a 2017 paper (de la Puente et al. 2017). The QIAseq Investigator Global Ancestry SNP panel was reported to have 164 SNPs in a 2021 paper. The QIAseq Investigator Missing Persons SNP panel or ICMP MPs-Plex panel was developed in

Table 1.2 Advancements in multiplexing of STR and SNP loci for NGS analysis for forensic applications

Year	NGS STR and SNP kits	Manufacturer	Number of loci
2014	HID-Ion Ampliseq™ Ancestry Panel	Applied Biosystems	165
2014	HID-Ion Ampliseq™ Identity Panel	Applied Biosystems	124
2016	PowerSeq™ 46GY	Promega	46
2018	Precision ID GlobalFiler™ NGS STR Panel	ThermoFisher	35
2017	QIAseq Investigator ID SNP panel	Qiagen	140
2017	ForenSeq™ Signature Prep Kit DPMA	Verogen	152
2017	ForenSeq™ Signature Prep Kit DPMB	Verogen	230
2021	QIAseq Investigator Global Ancestry SNP panel	Qiagen	164
2021	ForenSeq® MainstAY	Verogen	52
2021	ForenSeq® MainstAY SE	Verogen	53
2021	ForenSeq® Kintelligence Kit	Verogen	10,230
2022	ForenSeq® Imagen Kit DPME	Verogen	55
2022	ForenSeq® Imagen Kit DPMF	Verogen	111

Table 1.3 Advancements in mitochondrial DNA NGS kits for forensic applications

Year	Mitochondrial DNA NGS kit	Manufacturer
2018	Precision ID mtDNA Control Region Panel	Applied Biosystems
2018	Precision ID mtDNA Whole Genome Panel	Applied Biosystems
2020	ForenSeq™ mtDNA Control Region Kit	Verogen
2020	ForenSeq™ mtDNA Whole Genome Kit	Verogen
2020	QIAseq Investigator Human Mitochondria Control Region Panel	Qiagen
2021	PowerSeq® CRM Nested System	Promega

collaboration with the International Commission on Missing Persons (ICMP) to contain 1200 tri-allelic kinship SNPs, 34 X-SNPs, 55 aiSNPs, and 46 microhaplotypes as discussed at the International Symposium on Human Identification (ISHI) 2021 annual meeting.

NGS has also enabled single assay targeted mtDNA sequencing in the highly variable (HV) regions as wells as full mitochondrial chromosome sequencing (Table 1.3). Applied Biosystems, Qiagen, Promega, and Verogen now offer products for control region and whole genome mitochondrial sequencing. These options have greatly reduced the analysis time for HV or mtDNA analysis using Sanger sequencing or even a few 10-plex SNaPshot assays.

1.5 Addition of Genetic Genealogy to Solve Unsolved Cases

For years, the DNA typing process focused on analysis of a few to a couple dozen loci. The standard set of loci limited the analysis to sex determination and human identification. Additional loci and products are available for use in lineage typing. Advancements in forensic DNA typing have led to multiplexing of thousands of loci.

The additional loci have opened new avenues for analysis. As described above, one is genetic genealogy. Genetic genealogy is a new field that combines the information and individualization power from DNA typing with traditional historical genealogy research methods to trace distant relatives such as third cousins through a common ancestor. Tools such as GEDmatch compare DNA profiles obtained from evidence to DNA profiles obtained by living individuals through commercial personal genomics ancestry companies such as 23andMe, MyHeritage, AncestryDNA, LivingDNA, Findmypast, and FamilyTreeDNA. The first cold case solved with genetic genealogy was the "Golden State Killer" case solved in 2018 (Verogen 2020). As of this writing, according to publicly published reports, more than 100 cases have now been solved with investigative genetic genealogy (IGG) or forensic genetic genealogy (FGG), the use of direct-to-consumer genetic information to identify suspects or victims in criminal cases including decades-old cold cases (Dowdeswell 2022; Smirnova and Elkins 2023).

1.6 Conclusion

DNA analysis continues to change at a rapid pace. Even the experienced analyst may be surprised at the innovations and options. DNA has now been used in over 100 publicly reported cases in which genetic genealogy was used to solve the case. This next chapters in the book will detail the many advancements in DNA analysis from products to chemistry to technology to databases and crime solving.

References

Butler J (2005) Forensic DNA typing, 2nd edn. Elsevier Academic Press, Burlington, MA

DeAngelo TM, Elkins KM (2023) Ethical issues of forensic application of NGS technology. In: Dash HR, Elkins KM, Al-Snan NR (eds) NGS technology in DNA analysis. Elsevier. (in press)

de la Puente M, Phillips C, Santos C, Fondevila M, Carracedo Á, Lareu MV (2017) Evaluation of the Qiagen 140-SNP forensic identification multiplex for massively parallel sequencing. Forensic Sci Int Genet 28:35–43. https://doi.org/10.1016/j.fsigen.2017.01.012

Dowdeswell TL (2022) Forensic genetic genealogy: a profile of cases solved. Forensic Sci Int Genet 58:102679. https://doi.org/10.1016/j.fsigen.2022.102679

Edwards A, Civitello A, Hammond HA, Caskey CT (1991) DNA typing and genetic mapping with trimeric and tetrameric tandem repeats. Am J Hum Genet 49(4): 746–756

Elkins KM (2013) Forensic DNA biology: a laboratory manual. Elsevier Academic Press, Waltham, MA

Elkins KM, Zeller CB (2021) Next generation sequencing in forensic science: a primer. CRC Press, Boca Raton, FL

Elkins KM, Kollmann DD, Bowie P, Chart T, Gorr KK, Hutson B, Kiesel BD, Klavens A, Zeller CB (2022) Using DNA to develop a profile of phenotypic characteristics of historic period human remains. Forensic Sci Int Genet Suppl Ser 8:291–293

Ensenberger MG, Lenz KA, Matthies LK, Hadinoto GM, Schienman JE, Przech AJ, Morganti MW, Renstrom DT, Baker VM, Gawrys KM, Hoogendoorn M, Steffen CR, Martín P, Alonso A, Olson HR, Sprecher CJ, Storts DR (2016) Developmental validation of the PowerPlex(®) Fusion 6C System. Forensic Sci Int Genet. 21:134–144. https://doi.org/10.1016/j.fsigen.2015.12.011

Hares DR (2015) Selection and implementation of expanded CODIS core loci in the United States. Forensic Sci Int Genet 17:33–34. https://doi.org/10.1016/j.fsigen.2015.03.006

Jäger AC, Alvarez ML, Davis CP, Guzmán E, Han Y, Way L, Walichiewicz P, Silva D, Pham N, Caves G, Bruand J, Schlesinger F, Pond SJK, Varlaro J, Stephens KM, Holt CL (2017) Developmental validation of the MiSeq FGx forensic genomics system for targeted next generation sequencing in forensic DNA casework and database laboratories. Forensic Sci Int Genet 28:52–70. https://doi.org/10.1016/j.fsigen.2017.01.011

Liu Y, Guo L, Jin H, Li Z, Bai R, Shi M, Ma S (2017) Developmental validation of a 6-dye typing system with 27 loci and application in Han population of China. Sci Rep 7(1):4706. https://doi.org/10.1038/s41598-017-04548-1

Minogue TD, Koehler JW, Stefan CP, Conrad TA (2019) Next-generation sequencing for biodefense: Biothreat detection, forensics, and the clinic. Clin Chem 65(3): 383–392

Oldoni F, Kidd KK, Podini D (2019) Microhaplotypes in forensic genetics. Forensic Sci Int Genet 38:54–69. https://doi.org/10.1016/j.fsigen.2018.09.009

Patrick KL (2007) 454 Life sciences: illuminating the future of genome sequencing and personalized medicine. Yale J Biol Med 80(4):191–194

Rodriguez JJRB, Laude RP, De Ungria MCA (2021) An integrated system for forensic DNA testing of sexual assault cases in the Philippines. Forensic Sci Int Synerg. 3:100133. https://doi.org/10.1016/j.fsisyn.2021.100133

Scientific Working Group on DNA Analysis Methods. Interpretation guidelines for autosomal STR typing by forensic DNA testing laboratories. Approved 12 Jan 2017. https://1ecb9588-ea6f-4feb-971a-73265dbf079c.filesusr.com/ugd/4344b0_50e2749756a242528e6285a5bb478f4c.pdf. Accessed 23 Jan 2021

Scientific Working Group on DNA Analysis Methods. Addendum to "SWGDAM interpretation guidelines for autosomal STR typing by forensic DNA testing laboratories" to address next generation sequencing. Approved 23 Apr 2019. https://1ecb9588-ea6f-4feb-971a-73265dbf079c.filesusr.com/ugd/4344b0_91f2b89538844575a9f51867def7be85.pdf. Accessed 23 Jan 2021

Smirnova D, Elkins KM Application of next generation sequencing technology and investigative genetic genealogy to cold cases. In NGS technology in DNA analysis, Hirak Ranjan Dash, Kelly M Elkins, Noora R Al-Snan, Elsevier, 2023., in press

Turiello R, Nouwairi RL, Landers JP (2023) Taking the microfluidic approach to nucleic acid analysis in forensics: review and perspectives. Forensic Sci Int Genet:102824. https://doi.org/10.1016/j.fsigen.2022.102824

Verogen. FBI approves Verogen's next-gen forensic DNA technology for national DNA index system (NDIS). https://verogen.com/ndis-approval-of-miseq-fgx/. Accessed 7 Nov 2020

2 Use of Advanced Molecular Techniques for Human Body Fluids Detection

2.1 Introduction to Forensic Body Fluids Analysis

Forensic body fluids analysis is the oldest facet of forensic biology. Body fluids analysis has its roots in chemical color tests that were developed at the end of the nineteenth and beginning of the twentieth century. These include the acid phosphatase test for semen, Kastle–Meyer test for blood, and starch-iodine test for saliva (Virkler and Lednev 2009; Elkins 2013). Forensic scientists test for body fluids to identify samples that are biological and could be used for DNA typing and to determine the circumstances of a case. Perpetrators may insert objects to victims' body orifices, including objects that are not intended for that purpose. Identifying the presence of the body fluids on a sample can aid investigators in understanding the perpetrator's actions in the case. New biomolecule targets including small RNAs identified in disease research have been investigated for body fluids analysis (Courts and Madea 2010; Harbison and Fleming 2016; Lynch and Fleming 2021).

The focus of this chapter is advanced molecular techniques for human body fluids detection. Body fluids analysis has evolved and new methods have been developed that utilize next generation sequencing (NGS) approaches including pyrosequencing and massively parallel sequencing (MPS) (Børsting and Morling 2018). Extraction and quantification methods will be covered in Chap. 3 and technology will be covered in Chap. 7; the applications are the focus of this chapter. Pyrosequencing is being used to detect and differentiate body fluids by targeting epigenetic variation in DNA and MPS is being used to characterize the microbiome of different body fluid and body locations. Other advanced approaches including quantitative PCR are being used to differentiate body fluids by detecting and quantitating RNA molecules including messenger RNA (mRNA), microRNA, and small nuclear RNA (snRNA) expression. Analysis and differentiation of varied sample regions has been performed using 16S rRNA gene sequencing and PCR high resolution melt (HRM) epigenetic assays have been used to detect and identify body fluids. A major advantage of using RNA is that it can be coextracted with DNA from a small sample. The DNA can be used for DNA typing and human identification and the body fluids analysis can be performed on a sample that otherwise would be discarded.

2.2 mRNA Expression

In eukaryotes, cells differentiate early in embryonic development. The differentiated cells multiply to give rise to different organs and tissues with specialized functions. As a result of the tissue differentiation, cells in each tissue produce a set of proteins for their function. Each cell inherits a copy of the organism's genome in its nucleus (Elkins 2013). As explained by the

H. R. Dash et al., *Advancements in Forensic DNA Analysis*,
https://doi.org/10.1007/978-981-99-6195-5_2

Central Dogma of Molecular Biology, the DNA is transcribed to messenger RNA and the mRNA is translated to its protein. Researchers have discovered that the mRNA expression levels vary in different tissues and that this feature can be used to identify body fluids for forensic applications (Bauer and Patzelt 2002; Dørum et al. 2018; Hanson et al. 2018; Hussein et al. 2021). Several studies have focused on evaluating gene expression in different tissues and identifying mRNA levels that vary in the blood, semen, saliva, and vaginal fluid.

Bauer and Patzelt (2002) was as the first to apply mRNA markers to differentiate between menstrual fluid and vaginal mucosa. Hanson et al. (2018) used a 33-marker mRNA mRNA sequencing assay to correctly identify body fluids including semen, saliva, blood, vaginal secretions, menstrual blood, and skin. The RNA typing process has several steps and includes sampling, RNA extraction, reverse transcription (RT) to cDNA, and analysis by qPCR. DNA and RNA can be coextracted to isolate mRNA and other RNA molecules. This reduces the need for additional sample for RNA body fluids analysis testing. The expression of the target mRNAs is compared to housekeeping genes. Body fluids may express the same proteins but in different ratios. A reduced or enhanced expression can be used for differentiating the body fluids. The difference in threshold cycles (ΔCt) is compared between the gene of interest and the housekeeping gene reference. qPCR is extremely sensitive and can identify RNA molecules from a single cell (attomolar concentration). Menstrual blood, saliva, and nasal secretions have been differentiated using mRNA analysis using RT-PCR. (Hussein et al. 2021). The MMP7 and MMP10 markers were used to detect menstrual blood; the HTN3 and STATH markers were used to detect saliva; and BPIFA1 and STATH were used to detect nasal secretions. The mRNA levels decreased over time in the study from day one to day seven when stored at room temperature (Hussein et al. 2021). Recently, the origin of stains using mRNA was predicted using next generation sequencing (Zubakov et al. 2015; Dørum et al. 2018). Dørum et al. (2018) analyzed 183 body fluids/tissues and analyzed the NGS/MPS data by partial least squares (PLS) and linear discriminant analysis (LDA) (Dørum et al. 2018). The NGS read counts were used to classify the samples to one of six body fluids (Dørum et al. 2018). Ingold et al. (2020) reported on the success in a EUROFORGEN/EDNAP collaborative exercise in which two MPS assays were used for mRNA sequencing: Illumina MiSeq/FGx and Ion Torrent PGM/S5 employing an Illumina 33plex mRNA panel and a 35plex cSNP panel, respectively.

2.3 microRNA Expression

MicroRNAs (miRNA) (19–23 nt) are a category of small non-coding RNAs (El-Mogy et al. 2018). Their small size enables them to be detected after conditions that lead to degradation (Yang et al. 2014). These non-coding RNAs regulate metabolic processes including metastasis, proliferation and apoptosis and are mostly found in cells but also have been isolated from body fluids including saliva, tears, urine, plasma, serum, blood, seminal fluid, breast milk, amniotic fluid, and colostrum (El-Mogy et al. 2018). To date, the miRBase has catalogued 38,589 microRNAs (https://www.mirbase.org/).

MiRNAs have been investigated for forensic body fluids analysis (Hanson et al. 2009; Ye et al. 2019; Rocchi et al. 2020) and were proposed for analysis by NGS (Yang et al. 2014). Total RNA, including miRNAs, can be coeluted with DNA and tested using qPCR. Hanson et al. (2009) analyzed 452 miRNAs from blood, menstrual blood, saliva, semen, and vaginal secretions and analyzed them with real-time PCR. They observed differential expression of the markers in the body fluids at nine miRNAs (miR10b, miR16, miR124a, miR135b, miR451, miR205, miR372, miR412, and miR658) (Hanson et al. 2009). Quantitative RT-PCR Taq Man assays were developed for distinguishing body fluids including venous blood and semen and the expression of 718 miRNAs from semen, saliva, venous blood, menstrual blood, and vaginal secretions were analyzed using a microarray

(Zubakov et al. 2010). NGS using the Illumina MiSeq and Thermo Fisher Ion Torrent instruments has been employed for miRNA analysis and measuring differential expression. Several commercial kits are available including the QIAseq miRNA Library Kit (12 reactions) that includes the reagents for library preparation including reverse transcription, cDNA cleanup, and library cleanup and the HTG EdgeSeq miRNA whole transcriptome assay have been developed. The HTG EdgeSeq miRNA Whole Transcriptome Assay measures the expression of 2083 human miRNA transcripts directly from plasma or serum using NGS avoiding sample loss in DNA extraction steps. Control samples should be included in the experiments for quality control. Cloud-based and terminal-based bioinformatics analysis pipelines are available for users to analyze their data.

2.4 snRNA, snoRNA, and piRNA Expression

Small nuclear RNA (snRNA) molecules (150 bp) are abundant non-coding RNAs that were first identified in the 1970s. They have been shown to regulate transcription factors and are associated with specific proteins leading them to form small nuclear ribonuclear proteins (snRNPs). They are transcribed in the nucleus and aid in governing intron splicing and other cell processes. Many of these have been identified using NGS. Small nucleolar RNA (snoRNA) are small RNA molecules (60–170 nt) that guide chemical modifications of other RNAs including ribosomal RNA (rRNA), transfer RNA (tRNA), and snRNA molecules and impact cellular homeostasis and disease (Esteller 2011). Owing to their small sizes snRNAs and snoRNAs, are more resistant to degradation than mRNA molecules although they are longer than miRNAs (Yang et al. 2014; Liu et al. 2022). Piwi RNA (piRNA) are small RNAs (24–31 nt) that interact with PIWI proteins and suppress transposons and regulate genes (Iwasaki et al. 2015). Many small RNAs have been studied in cancer research and shown to play roles in tumorigenesis (Vallone et al. 2018). SnoRNAs, snRNAs, piRNAs, and miRNAs can also be used for forensic body fluids analysis and were investigated using 30 body fluid samples (saliva, semen, menstrual blood, venous blood, vaginal secretion, and skin site) in a recent study (Liu et al. 2022). Relative and absolute specific markers were defined (Liu et al. 2022). The loci investigated included 1394 miRNAs, 636 snoRNAs, 3157 piRNAs, and 17 snRNAs (Liu et al. 2022). Variations in small RNA expressions were observed across the sample types and 90 absolute specific markers were screened for specific expression in a body fluid (Liu et al. 2022).

2.5 Methylated DNA Analysis

Methylation, or addition of a methyl group to cytosine bases in DNA strands in CpG islands, is an epigenetic modification to DNA (Lilischkis et al. 2001; Bernstein et al. 2007). Tissue differentially methylated regions (tDMRs) (Weber et al. 2005) have been explored for body fluid differentiation (Rakyan et al. 2008) and assays including pyrosequencing, MPS, and PCR HRM have been developed for forensic applications (Lee et al. 2012). In 2010, Paliwal et al. (2010) reported on the quantitative determination of methylation levels in “minute” quantities of body fluids. In 2011, DNA methylation was probed using a methylation-sensitive restriction endonuclease and fluorescent-labeled primers were used for multiplex PCR (Frumkin et al. 2011). The amplified products were separated using capillary electrophoresis (CE) (Frumkin et al. 2011). Fifty blood, semen, saliva, and skin epidermis samples were tested and the assay worked successfully in all of the samples (Frumkin et al. 2011). The assay was also able to detect semen and saliva in various ratios and correctly predict the percentage of semen (Frumkin et al. 2011). Bisulfite genomic sequencing was conducted on DNA extracted from a dried blood spot (Xu et al. 2012). Slieker et al. (2013) and Park et al. (2014) applied Illumina 450k chip data to identifying body fluid specific CpG sites. Slieker et al. (2013) found that tDMRs are associated with CpG-poor regions and alternate transcription sites for the different tissues. Park et al. (2014) published their results probing over 450,000 CpG sites in which they identified the novel CpG loci: "cg06379435 and cg08792630 for blood, cg26107890 and cg20691722 for saliva, cg23521140 and cg17610929 for semen, and cg01774894 and cg14991487 for vaginal secretions."

Methods for creating primers for pyrosequencing assays for body fluids analysis have been published (Qiagen 2021; Elkins 2021). In tDMR sequencing, pyrosequencing, and PCR HRM assays, DNA is extracted from body fluid samples and bisulfite treated prior to testing (Grunau et al. 2001; Meissner et al. 2005; Dejeux et al. 2009). In 2012, Madi et al. (2012) used pyrosequencing to differentiate body fluids: ZC3H12D and FGF7 were found to differentiate sperm from blood, saliva, and epithelial cells; C20orf117 was found to differentiate blood from saliva, sperm, and epithelial cells; and BCAS4 was found to differentiate saliva from sperm, blood, and epithelial cells. Sperm were identified by tDMR methylation analysis by pyrosequencing following bisulfite modification in a subsequent study (Balamurugan et al. 2014) and vaginal epithelia was discriminated by pyrosequencing by Antunes et al. (2016b). Using PCR HRM, Antunes et al. (2016a) used the Zinc Finger CCCH-Type Containing 12D (ZC3H12D) locus to differentiate hypomethylated semen and hypermethylated blood and saliva. Levine and Zeller (2018) reported on the use of PCR HRM assays to differentiate semen from other body fluids by differences in methylation using the loci ZC3H12D and DACT1. Powell et al. (2018) further reported on the the identification of sperm samples and differentiation from saliva samples using five ZC3H12D CpG positions via pyrosequencing; the saliva sample exhibited significantly higher methylation than the sperm sample but aspermic semen exhibited a comparable methylation level as the saliva sample. The results were congruent with previous findings by Silva et al. (2016) on semen, saliva, and blood samples using the markers ZC3H12D, BCAS4 and cg06379435, respectively. Alghanim et al.

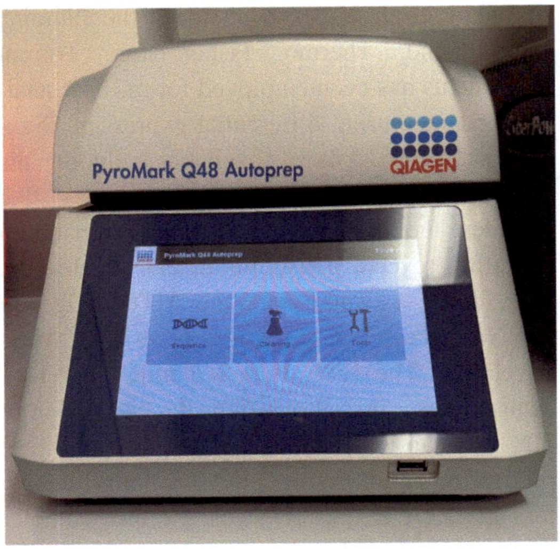

Fig. 2.1 Qiagen Q48 Pyrosequencer Instrument

(2020) reported that they developed four epigenetic pyrosequencing and PCR HRM assays for body fluids including saliva, sperm, semen without sperm, sweat/skin, and venous and menstrual blood. Pyrosequencing assays were developed using the AHRR, NMUR2, UBE2U, and SA-6 markers and PCR HRM assays were developed using the NMUR2, UBE2U, and SA-6 markers (Alghanim et al. 2020). The NMUR2 and UBE2U markers were found to be specific for sperm which has lower DNA methylation levels than the other body fluids (Alghanim et al. 2020). The SA-6 marker was found to be hypermethylated in saliva as compared to the other body fluids (Alghanim et al. 2020). In a subsequent study, the group evaluated the relative methylation level for each body fluid (2 blood markers, 1 vaginal marker, and 1 buccal marker) using qPCR HRM (Antunes et al. 2021). Additional PCR HRM assays to differentiate single and mixed body fluids have been developed (Zeller, Levine, Zeidman, Hay, and Ayers, unpublished results). The CpG sites in the AHRR locus were found to be hypomethylated in blood as compared to the other body fluids (Alghanim et al. 2020). In another study marker, the SPERM2 marker was found to be better than the SEU2 marker for the identification of semen in mixed samples but the SEU2 performed well in unmixed sample (Gomaa et al. 2021). BLM1 and

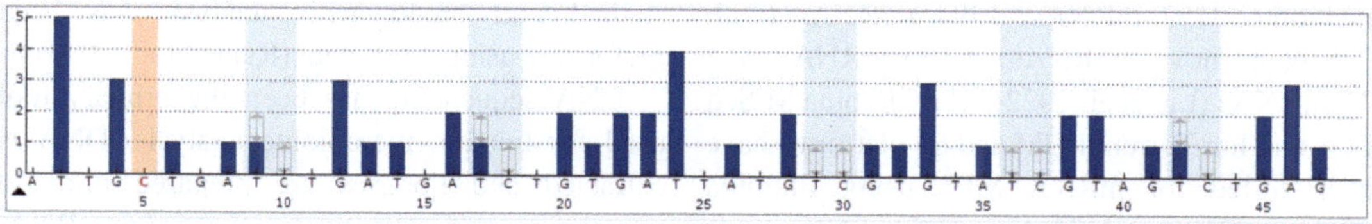

Fig. 2.2 Qiagen Q48 tDMR pyrosequencing assay for blood; the orange shaded base addition is a negative control and the gray shaded bases presence/absence are probed in the assay

MENS1 markers identified peripheral and menstrual blood, respectively (Gomaa et al. 2021). The markers WDR11, PON2, and NHSL1 were found to have lower melt temperatures in PCR HRM in sweat/skin epithelial cells as compared to body fluids blood, saliva, vaginal epithelia, and semen (Fernandez-Tejero et al. 2022). A commercial pyrosequencing assay for body fluids was developed by the McCord group (Powell et al. 2018, Gauthier et al. 2019) and licensed by Qiagen and can be run on the Qiagen Q48 (Fig. 2.1) (Qiagen 2020). A sample pyrogram is shown in Fig. 2.2. The AllPrep DNA/RNA/miRNA Universal Kit was evaluated in 2020 (Watanabe and Akutsu 2020). RNA and DNA was extracted from blood and saliva stains and pyrosequencing results revealed that the methylation ratios of DNA samples extracted by the universal co-extraction kit were not different from those extracted by other DNA extraction kits thereby streamlining the body fluid identification process (Watanabe and Akutsu 2020).

Gene methylation in testis, prostrate, ovaries, cervix, uterus, blood, and salivary gland tissues were identified from genome-wide expression datasets, NGSmethDB and ENCODE. After evaluating 53 potential genes, ZNF282 and HPCAL1 were identified to be preferentially expressed in the cervix. The ZNF282 and HPCAL1 CpGs islands were hypomethylated in semen and saliva, respectively. Under environmental conditions in which samples were dried, left outside, stored in moist conditions, and sprayed with bleach and storage for 50 days, the saliva sample outside "on the ground" had a significant decrease in methylation level while there was no change for vaginal fluid in all of the conditions and no differences in methylation levels for vaginal fluid were observed after 50 days (Ghai et al. 2020).

A multiplex assay that probed nine CpG markers to differentiate five body fluids (blood, semen, saliva, vaginal fluid, and menstrual blood) was tested using single base extension (SBE) with 22 samples from 18 cases previously tested with acid phosphatase for semen, leucomalachite green for blood, and SALIgAE® for saliva. All of the samples were presumptive positive except one which displayed an all-negative result. The test was especially useful for vaginal fluid and menstrual blood for which there is no presumptive test (Choung et al. 2021).

Another assay used amplification refractory mutation system-PCR (ARMS-PCR) to amplify DNA from five body fluids including saliva, semen, venous blood, menstrual blood, and vaginal fluid using multiplexed methylation reactions with 22 markers. Following PCR, CE separated and detected the products and a random forest model was used to detect DNA methylation patterns and identify the body fluids (Tian et al. 2020).

2.6 MPS DNA Analysis

MPS is the newest approach to body fluids analysis (Ingold et al. 2020; Liu et al. 2022; Watanabe et al. 2022). Yang et al. proposed NGS for miRNA analysis in 2014 (Yang et al. 2014). In a study of 30 forensically-relevant samples including body fluids saliva, semen, vaginal fluid, venous and menstrual blood and skin, miRNA, piRNA, snoRNA, and snRNA markers were sequenced (Liu et al. 2022). This led to the discovery of 5204 small RNAs including 3157 piRNAs, 1394 miRNAs (with 236 being novel miRNAs), 636 snoRNAs, and 17 snRNAs (Liu et al. 2022). The Illumina MiSeq FGx and Ion Torrent PGM and S5 platforms have been tested for mRNA MPS using Illumina 33plex mRNA and 35plex cDNA SNP panels (Ingold et al. 2020). The authors reported that most of the single source fluid samples and some of the mixture components were identified by a partial least squares discriminant analysis (PLS-DA) model (Ingold et al. 2020). As an alternative to α-amylase activity, MPS was used to analyze five saliva-specific SNPs in an unmethylated region and 13 in a methylated region in a genome-wide study (Watanabe et al. 2022). Following bisulfite treatment, the saliva samples and mixtures of body fluids including saliva, blood, semen, and vaginal fluid were sequenced and the sequences were found to be useful for individualization and determination of the body fluid (Watanabe et al. 2022). Zubakov et al. (2015) published a proof-of-principle study of combined individual and tissue identification by using an Ion Torrent PGM to sequence STRs, amelogenin, and mRNAs. In another study, the origin of stains was predicted using MPS and the miRNome (Dørum et al. 2019).

2.7 Microbiome Analysis for Body Fluids and Sites

The human microbiome was the focus of a 10-year U.S. National Institutes of Health (NIH) Human Microbiome Project (HMP) study that spanned 2007–2016 (Belizário and Napolitano 2015). Adults and infants were both studied. Samples were collected from participants' skin, oral, nasal, gastrointestinal (GI), and urogenital regions (Belizário and Napolitano 2015) and vaginal, skin, stool, and oral cavity (Pasolli et al. 2019). The skin, GI, nasal, and oral mucosa regions were sampled in infants (Plummer et al. 2015; Milani et al. 2017). Oral microbiomes have been investigated in several studies (Guranlian 2007; Siqueira et al. 2012; Chen et al. 2014; Hasan et al. 2014; Kistler et al. 2015; Willis and Gabaldón 2020). Fecal samples have also been studied and phylogenetic analysis has been performed using NGS data (Rajan et al. 2019).

DNA sequencing has been the primary tool used for human microbiome analysis. DNA is extracted and libraries are prepared prior to sequencing and data analysis. As with RNA, microbial DNA can be coextracted with human DNA (Alessandrini et al. 2019). Whole genome sequencing has been conducted using an Illumina HiSeq NGS instrument (Pasolli et al. 2019).

Primers targeting the 16S rRNA gene of the small 30S prokaryotic ribosomal subunit have also been used. The 16S rRNA gene is highly conserved and widely used for taxonomic analysis and determining relative proportions of bacterial taxa (Guranlian 2007; Kistler et al. 2015; Franco-Duarte et al. 2019). The 16S rRNA gene contains nine variable (V) regions (with the ranges based on the *E. coli* genome numbering: V1, 69–99; V2, 137–242; V3, 433–497; V4, 576–682; V5, 822–879; V6: 986–1043; V7, 1117–1173; V8, 1243–1294; and V9, 1435–1465) (Chakravorty et al. 2007). Whereas Sanger or SNaPshot sequencing would focus on a few SNPs or one of the variable regions at a time, all of the regions can be sequenced simultaneously and more efficiently using NGS without targeting a specific region (Cho et al. 2019).

In several studies, sequence data was analyzed using taxonomic profiling and the bacterial composition and abundance was found to differ among the body regions tested. The 16S rRNA gene was probed to analyze intestinal microbiota using SOLiD shotgun sequencing (Mitra et al. 2013). Sequencing the 16S rRNA region elucidated variations in three oral microbiomes (Siqueira et al. 2012). In another study, the bacterial composition varied across the vaginal, skin, stool, and oral cavity locations (Pasolli et al. 2019; Chowdhury and Fong 2020) Saliva bacteria from two HMP samples was characterized by NGS and species composition was analyzed by Hasan et al. (2014) and the skin, GI, nasal, and oral mucosa regions were sampled in infants (Milani et al. 2017).

In addition, NGS was used to probe the microbiome of 18 samples including vaginal, oral, and fecal samples as well as a yogurt distractor sample. Microbiome analysis led to the identification of each species in the samples and identify vaginal fluid from the other samples (Giampaoli et al. 2017). In a separate study, NGS characterized bacteria and viruses in urine and determined that it was not bacteria-free (Wojciuk et al. 2019).

The cell type from body sites including hand, foot, groin, penis, vagina, mouth, and feces were analyzed (Quaak et al. 2018). The oral and fecal sites could be distinguished from the skin and vaginal sites. Human feces were differentiable from dog and cat animal feces by analyzing the skin cell types. However, caution must be applied with differentiating samples; some skin site and penis site samples were highly similar to vaginal and skin site samples (Quaak et al. 2018). A study of three body sites for 51 individuals indicated that the microbiome varied more between individuals than within an individual (Woerner et al. 2019). Another study on the skin microbiome revealed that variations between individuals are greater than within an individual but caution must be used as the profile can change over time due to changes in diet, environment, or health condition (Tozzo et al. 2020).

The hidSkinPlex tool was developed from human microbiome research for forensic identification and body site prediction (Schmedes et al. 2018). Developmental testing was conducted using three skin locations: hand, chest, and foot. Body site prediction using the microbiome was found to be up to 86% accurate (Schmedes et al. 2018). In sum, these studies show that microbiome analysis can indicate which body location a sample originated from. Analysis of the urobiome and vaginal microbiome could aid in sexual assault investigations in forensic cases (Elkins and Zeller 2021).

In addition, study results have shown that characteristic microbiome taxonomy was associated with externally visible traits including obesity, psoriasis, acne, skin cancer, and periodontal disease (Guranlian 2007; Chen et al. 2014; Belizário and Napolitano 2015; Kistler et al. 2015). Microbiome testing of human hairs was found to differentiate male and female donors (Tridico et al. 2014). Microbiome analysis offers many case analysis opportunities for forensic scientists (Kuiper 2016). However, there are still challenges to wider adoption of microbiome testing for forensic applications. These include a lack of standardized analysis standardized operating procedures (SOPs) and tools, lack of user-friendly analysis tools, time for library prep and sequencing, sequencing bias and depth, reproducibility, handling large data storage, databases for analysis, and acceptance by courts (Budowle et al. 2014; Elkins and Zeller 2021). Additionally, sequence data must be analyzed carefully as some species may have multiple copies of the 16S rRNA gene and high numbers of microorganisms may be present in a sample while others may have only one copy (Sangiovanni et al. 2019) be present in low quantity which could be detected only if lower read counts are retained; this approach makes detecting contamination more likely (Sangiovanni et al. 2019).

2.8 Conclusion and Future of Body Fluids Analysis

Small RNAs, epigenetic modifications to DNA, and the microbiome have been shown to be useful targets for advanced analysis of body fluids and human body sites. While color tests from the last century or earlier can be used to presumptively identify body fluids including blood, semen, urine, feces, and saliva quickly and cheaply, they use more sample because the tests are separate. RNA and DNA targets can be used to detect and identify multiple body fluids simultaneously in one pyrosequencing, qPCR, PCR HRM, or NGS assay. Other drawbacks of the color tests are that there are no tests available to detect vaginal fluid or to differentiate between venous and menstrual blood. Vaginal fluid, menstrual and peripheral blood, semen, sperm, sweat/skin, and buccal cells/saliva can be differentiated by pyrosequencing and PCR HRM. RNA testing can detect and identify semen, blood, vaginal fluid, and saliva. Additional fluids including tears, urine, urogenital region, and feces can be identified using NGS and microbiome analysis.

Future research directions include continuing to evaluate new markers including specific, absolute, and reference markers. Further investigations into additional small RNAs including snoRNAs, snRNAs, mRNAs, microRNAs, as well as piwi-interacting RNAs may lead to the development of additional assays for molecular diagnostics for body fluid identification.

References

Alessandrini F, Brenciani A, Fioriti S, Melchionda F, Mingoia M, Morroni G, Tagliabracci A (2019) Validation of a universal DNA extraction method for human and microbiAL DNA analysis. Forensic Sci Int Genet Suppl Ser 7(1):256–258. https://doi.org/10.1016/j.fsigss.2019.09.098

Alghanim H, Balamurugan K, McCord B (2020) Development of DNA methylation markers for sperm, saliva and blood identification using pyrosequencing and qPCR/HRM. Anal Biochem 611:113933. https://doi.org/10.1016/j.ab.2020.113933

Antunes J, Silva DSBS, Balamurugan K, Duncan G, Alho CS, McCord B (2016a) High-resolution melt analysis of DNA methylation to discriminate semen in biological stains. Anal Biochem 494:40–45. https://doi.org/10.1016/j.ab.2015.10.002.10

Antunes J, Silva DS, Balamurugan K, Duncan G, Alho CS, McCord B (2016b) Forensic discrimination of vaginal epithelia by DNA methylation analysis through pyrosequencing. Electrophoresis 37(21):2751–2758. https://doi.org/10.1002/elps.201600037

Antunes J, Gauthier Q, Aguiar-Pulido V, Duncan G, McCord B (2021) A data-driven, high-throughput methodology to determine tissue-specific differentially methylated regions able to discriminate body fluids. Electrophoresis 42:1168–1176. https://doi.org/10.1002/elps.202000217.7

Balamurugan K, Bombardi R, Duncan G, McCord B (2014) Identification of spermatozoa by tissue-specific differential DNA methylation using bisulfite modification and pyrosequencing. Electrophoresis 35(21–22):3079–3086. https://doi.org/10.1002/elps.201400175

Bauer M, Patzelt D (2002) Evaluation of mRNA markers for the identification of menstrual blood. J Forensic Sci 47(6):1278–1282. PMID: 12455649

Belizário JE, Napolitano M (2015) Human microbiomes and their roles in dysbiosis, common diseases, and novel therapeutic approaches. Front Microbiol 6: 1050. https://doi.org/10.3389/fmicb.2015.01050

Bernstein BE, Meissner A, Lander ES (2007) The mammalian epigenome. Cell 128(4):669–681. https://doi.org/10.1016/j.cell.2007.01.033

Børsting C, Morling N (2018) Next generation sequencing and its applications in forensic genetics. Forensic Sci Int Genet 18:78–89. https://doi.org/10.1016/j.fsigen.2015.02.002

Budowle B, Connell ND, Bielecka-Oder A, Colwell RR, Corbett CR, Fletcher J, Forsman M, Kadavy DR, Markotic A, Morse SA, Murch RS, Sajantila A, Schmedes SE, Ternus KL, Turner SD, Minot S (2014) Validation of high throughput sequencing and microbial forensics applications. Investig Genet 5:9. https://doi.org/10.1186/2041-2223-5-9

Chakravorty S, Helb D, Burday M, Connell N, Alland D (2007) A detailed analysis of 16S ribosomal RNA gene segments for the diagnosis of pathogenic bacteria. J Microbiol Methods 69(2):330–339. https://doi.org/10.1016/j.mimet.2007.02.005

Chen W, Cheng YM, Zhang SW, Pan Q (2014) Supervised method for periodontitis phenotypes prediction based on microbial composition using 16S rRNA sequences. Int J Comput Biol Drug Des 7(2):214–224. https://doi.org/10.1504/IJCBDD.2014.061647

Cho Y, Lee MH, Kim HS, Park M, Kim M-H, Kwon H, Kim J, Lee YH, Lee DS (2019) Comparative analysis of Sanger and next generation sequencing methods for 16S rDNA analysis of post-mortem specimens. Aust J Forensic Sci 51(5):426–455. https://doi.org/10.1080/00450618.2017.1402957

Choung CM, Lee JW, Park JH, Kim CH, Park HC, Lim SK (2021) A forensic case study for body fluid identification using DNA methylation analysis. Legal Med 51:101872. https://doi.org/10.1016/j.legalmed.2021.101872

Chowdhury S, Fong SS (2020) Computational modeling of the human microbiome. Microorganisms 8(2):197. https://doi.org/10.3390/microorganisms8020197

Courts C, Madea B (2010) Micro-RNA—a potential for forensic science? Forensic Sci Int 203(1–3):106–111. https://doi.org/10.1016/j.forsciint.2010.07.002

Dejeux E, El Abdalaoui H, Gut IG, Tost J (2009) Identification and quantification of differentially methylated loci by pyrosequencing™ technology. In: Tost J (ed) Methods in molecular biology: DNA methylation: methods and protocols, vol 507, 2nd edn. Humana Press, New York, pp 189–205. https://doi.org/10.1007/978-1-59745-522-0_15

Dørum G, Ingold S, Hanson E, Ballantyne J, Snipen L, Haas C (2018) Predicting the origin of stains from next generation sequencing mRNA data. Forensic Sci Int Genet 34:37–48. https://doi.org/10.1016/j.fsigen.2018.01.001

Dørum G, Ingold S, Hanson E, Ballantyne J, Russo G, Aluri S, Snipen L, Haas C (2019) Predicting the origin of stains from whole miRNome massively parallel sequencing data. Forensic Sci Int Genet 40:131–139. https://doi.org/10.1016/j.fsigen.2019.02.015

Elkins KM (2013) Forensic DNA biology: a laboratory manual. Elsevier Academic Press, Waltham, MA

Elkins KM (2021) Pyrosequencing primer design for forensic biology applications. In: Basu C (ed) Methods in molecular biology: PCR primer design, 3rd edn. Humana Press, New York. https://doi.org/10.1007/978-1-0716-1799-1

Elkins KM, Zeller CB (2021) Next generation sequencing in forensic science: a primer. Taylor & Francis/CRC Press, Boca Raton, FL

El-Mogy M, Lam B, Haj-Ahmad TA, McGowan S, Yu D, Nosal L, Rghei N, Roberts P, Haj-Ahmad Y (2018) Diversity and signature of small RNA in different bodily fluids using next generation sequencing. BMC Genomics 19(1):1–24

Esteller M (2011) Non-coding RNAs in human disease. Nat Rev Genet 12:861–874

Fernandez-Tejero N, Gauthier Q, Cho S, McCord BR (2022) High-resolution melt analysis for the detection of skin/sweat via DNA methylation. Electrophoresis 44:371–377. https://doi.org/10.1002/elps.202200120.6

Franco-Duarte R, Černáková L, Snehal K, Kaushik KS, Salehi B, Bevilacqua A, Corbo MR, Antolak H, Dybka-Stępień K, Leszczewicz M et al (2019) Advances in chemical and biological methods to identify microorganisms—from past to present. Microorganisms 7(5):130. https://doi.org/10.3390/microorganisms7050130

Frumkin D, Wasserstrom A, Budowle B, Davidson A (2011) DNA methylation-based forensic tissue identification. Forensic Sci Int Genet 5(5):517–524. https://doi.org/10.1016/j.fsigen.2010.12.001

Gauthier QT, Cho S, Carmel JH, McCord BR (2019) Development of a body fluid identification multiplex via DNA methylation analysis. Electrophoresis 40(18–19):2565–2574. https://doi.org/10.1002/elps.201900118

Ghai M, Naidoo N, Evans DL, Kader F (2020) Identification of novel semen and saliva specific methylation markers and its potential application in forensic analysis. Forensic Sci Int Genet 49:102392. https://doi.org/10.1016/j.fsigen.2020.102392

Giampaoli S, DeVittori E, Valeriani F, Berti A, Spica VR (2017) Informativeness of NGS analysis for vaginal fluid identification. J Forensic Sci 62(1):192–196. https://doi.org/10.1111/1556-4029.13222

Gomaa R, Nader L, Jamal J (2021) Application of DNA methylation-based markers in identification of mixed body fluid evidences simulation crime scene scenarios. Egypt J Forensic Sci 11:12. https://doi.org/10.1186/s41935-021-00226-6

Grunau C, Clark SJ, Rosenthal A (2001) Bisulfite genomic sequencing: systematic investigation of critical experimental parameters. Nucleic Acids Res 29(13):E65. https://doi.org/10.1093/nar/29.13.e65

Guranlian JR (2007) The role of dental plaque biofilm in oral health. J Dent Hyg 81(5):116

Hanson EK, Lubenow H, Ballantyne J (2009) Identification of forensically relevant body fluids using a panel of differentially expressed microRNAs. Anal Biochem 387(2):303–314. https://doi.org/10.1016/j.ab.2009.01.037

Hanson E, Ingold S, Haas C, Ballantyne J (2018) Messenger RNA biomarker signatures for forensic body fluid identification revealed by targeted RNA sequencing. Forensic Sci Int Genet 34:206–221. https://doi.org/10.1016/j.fsigen.2018.02.020

Harbison S, Fleming R (2016) Forensic body fluid identification: state of the art. Res Rep Forensic Med Sci 6(2016):11–23

Hasan NA, Young BA, Minard-Smith AT, Saeed K, Li H, Heizer EM, McMillan NJ, Isom R, Abdullah AS, Bornman DM (2014) Microbial community profiling of human saliva using shotgun metagenomic sequencing. PLoS One 9:e97699. https://doi.org/10.1371/journal.pone.0097699

Hussein RF, El Mahdy SM, Saged NM, Rashed L, Ghaleb SS (2021) Identification of body fluids—menstrual blood, saliva, and nasal secretions—over different periods of time, using mRNA. Egypt J Forensic Sci 11(1):1–6

Ingold S, Dorum G, Hanson E, Ballard D, Berti A, Gettings KB et al (2020) Body fluid identification and assignment to donorrs using a targeted mRNA massively parallel sequencing approach—results of a second EUROFORGEN/EDNAP collaborative exercise. Forensic Sci Int Genet 45:102208. https://doi.org/10.1016/j.fsigen.2019.102208

Iwasaki YW, Siomi MC, Siomi H (2015) PIWI-interacting RNA: its biogenesis and functions. Annu Rev Biochem 84:405–433. https://doi.org/10.1146/annurev-biochem-060614-034258

Kistler JO, Pesaro M, Wade WG (2015) Development and pyrosequencing analysis of an in-vitro oral biofilm model. BMC Microbiol 15:24. https://doi.org/10.1186/s12866-015-0364-1

Kuiper I (2016) Microbial forensics: next-generation sequencing as catalyst: the use of new sequencing technologies to analyze whole microbial communities could become a powerful tool for forensic and criminal investigations. EMBO Rep 17:1085–1087. https://doi.org/10.15252/embr.201642794

Lee HY, Park MJ, Choi A, An JH, Yang WI, Shin K-J (2012) Potential forensic application of DNA methylation profiling to body fluid identification. Int J Legal Med 126:55–62

Levin R, Zeller CB (2018) The identification of forensically relevant body fluids using methylation-specific polymerase chain reaction (PCR) and high-resolution melt (HRM) analysis. Proceedings of the American academy of forensic sciences 70th annual scientific meeting, pp 313

Lilischkis R, Kneitz H, Kreipe H (2001) Methylation analysis of CpG islands. In: Brooks SA, Schumacher U (eds) Methods in molecular medicine: metastasis research protocols, vol I: cells and tissues, vol 57. Humana Press, New York, pp 271–283. https://doi.org/10.1385/1-59259-136-1:271

Liu Z, Wang Q, Wang N, Zang Y, Wu R, Sun H (2022) A comprehensive characterization of small RNA profiles by massively parallel sequencing in six forensic body fluids/tissue. Genes (Basel) 13(9):1530. https://doi.org/10.3390/genes13091530

Lynch C, Fleming R (2021) RNA-based approaches for body fluid identification in forensic science. Wiley Interdiscip Rev Forensic Sci 3(4):e1407. https://doi.org/10.1002/wfs2.1407

Madi T, Balamurugan K, Bombardi R, Duncan G, McCord B (2012) The determination of tissue-specific DNA methylation patterns in forensic biofluids using bisulfite modification and pyrosequencing. Electrophoresis 33:1736–1745. https://doi.org/10.1002/elps.201100711

Meissner A, Gnirke A, Bell GW, Ramsahoye B, Lander ES, Jaenisch R (2005) Reduced representation bisulfite sequencing for comparative high-resolution DNA methylation analysis. Nucleic Acids Res 33(18):5868–5877. https://doi.org/10.1093/nar/gki901

Milani C, Duranti S, Bottacini F, Casey E, Turroni F, Mahony J, Belzer C, Palacio SD, Montes SA, Mancabelli L et al (2017) The first microbial colonizers of the human gut: composition, activities, and health implications of the infant gut microbiota. Microbiol Mol Biol Rev 81(4):e00036-17. https://doi.org/10.1128/MMBR.00036-17

Mitra S, Förster-Fromme K, Damms-Machado A, Scheurenbrand T, Biskup S, Huson DH, Bischoff SC (2013) Analysis of the intestinal microbiota using SOLiD 16S rRNA gene sequencing and SOLiD shotgun sequencing. BMC Genomics 14(Suppl. 5):S16. https://doi.org/10.1186/1471-2164-14-S5-S16

Paliwal A, Vaissiere T, Herceg Z (2010) Quantitative detection of DNA methylation states in minute amounts of DNA from body fluids. Methods 52(3):242–247. https://doi.org/10.1016/j.ymeth.2010.03.008

Park J-L, Kwon O-H, Kim JH, Yoo H-S, Lee H-C, Woo K-M, Kim S-Y, Lee S-H, Kim YS (2014) Identification of body fluid-specific DNA methylation markers for use in forensic science. Forensic Sci Int Genet 13:147–153. https://doi.org/10.1016/j.fsigen.2014.07.011

Pasolli E, Asnicar F, Manara S, Zolfo M, Karcher N, Armanini F, Begini F, Manghi P, Tett A, Ghensi P, Collado MC, Rice BL, DuLong C, Morgan XC, Golden CD, Quince C, Huttenhower C, Segata N (2019) Extensive unexplored human microbiome diversity revealed by over 150,000 genomes from metagenomes spanning age, geography, and lifestyle. Cell 176(3):649–662. https://doi.org/10.1016/j.cell.2019.01.001

Plummer E, Twin J, Bulach DM, Garland SM, Tabrizi SN (2015) A comparison of three bioinformatics pipelines for the analysis of preterm gut microbiota using 16S rRNA gene sequencing data. J Proteomics Bioinformatics 8(12):12. https://doi.org/10.4172/jpb.1000381

Powell M, Lee AS, St. Andre P, McCord B (2018) Tissue source attribution using the PyroMark® Q48 autoprep system: sperm identification in forensic casework. Qiagen Applications Note. https://www.qiagen.com/us/resources/download.aspx?id=ddaa262e-f3ec-4ac7-9bac-aaf3ec8968cf&lang=en. Accessed 11 Jan 2021

Qiagen. PyroMark® Q48 autoprep user manual, June 2020. https://www.qiagen.com/us/resources/download.aspx?id=650a0c13-3b8e-4a77-b433-6b1e50b9525a&lang=en. Accessed 11 Jan 2021

Qiagen. PyroMark assay design SW 2.0 quick-start guide—(EN). https://www.qiagen.com/us/resources/download.aspx?id=231a6894-57d4-4f0b-81f2-eaa56a2b6bd8&lang=en. Accessed 11 Jan 2021

Quaak FCA, van Duijn T, Hoogenboom J, Kloosterman AD, Kuiper I (2018) Human-associated microbial populations as evidence in forensic casework. Forensic Sci Int Genet 36:176–185. https://doi.org/10.1016/j.fsigen.2018.06.020

Rajan SK, Lindqvist M, Brummer RJ, Schoultz I, Repsilber D (2019) Phylogenetic microbiota profiling in fecal samples depends on combination of sequencing depth and choice of NGS analysis method. PLoS One 14(9):e0222171. https://doi.org/10.1371/journal.pone.0222171

Rakyan VK, Down TA, Thorne NP, Flicek P, Kulesha E, Graf S, Tomazou EM, Bäckdahl L, Johnson N, Herberth M, Howe KL, Jackson DK, Miretti MM, Fiegler H, Marioni JC, Birney E, Hubbard TJ, Carter NP, Tavaré S, Beck S (2008) An integrated resource for genome-wide identification and analysis of human tissue-specific differentially methylated regions (tDMRs). Genome Res 18(9):1518–1529. https://doi.org/10.1101/gr.077479.108

Rocchi A, Chiti E, Maiese A, Turillazzi E, Spinetti I (2020) MicroRNAs: an update of applications in forensic science. Diagnostics 11(1):32. https://doi.org/10.1186/s12864-018-4785-8

Sangiovanni M, Granata I, Thind AS, Guarracino MR (2019) From trash to treasure: detecting unexpected contamination in unmapped NGS data. BMC Bioinformatics 20(Suppl. 4):168. https://doi.org/10.1186/s12859-019-2684-x

Schmedes SE, Woerner AE, Novroski NMM, Wendt FR, King JL, Stephens KM, Budowle B (2018) Targeted sequencing of clade-specific markers from skin microbiomes for forensic human identification. Forensic Sci Int Genet 32:50–61. https://doi.org/10.1016/j.fsigen.2017.10.004

Silva DSBS, Antunes J, Balamurugan K, Duncan G, Alho CS, McCord B (2016) Developmental validation studies of epigenetic DNA methylation markers for the detection of blood, semen and saliva samples. Forensic Sci Int Genet 23:55–63. https://doi.org/10.1016/j.fsigen.2016.01.017

Siqueira JF, Fouad AF, Rôças IN (2012) Pyrosequencing as a tool for better understanding of human microbiomes. J Oral Microbiol 4(1):10743. https://doi.org/10.3402/jom.v4i0.10743

Slieker RC, Bos SD, Goeman JJ, Bovée JV, Talens RP, van der Breggen R, Suchiman HED, Lameijer E-W, Putter H, van den Akker EB, Zhang Y, Jukema JW, Slagboom PE, Meulenbelt I, Heijmans BT (2013) Identification and systematic annotation of tissue-specific differentially methylated regions using the

Illumina 450k array. Epigenet Chromatin 6(1):26. https://doi.org/10.1186/1756-8935-6-26

Tian H, Bai P, Tan Y, Li Z, Peng D, Xiao X et al (2020) A new method to detect methylation profiles for forensic body fluid identification combining AMS-PC technique and random forest model. Forensic Sci Int Genet 49:102371. https://doi.org/10.1016/j.fsigen.2020.102371

Tozzo P, D'Angiolella G, Brun P, Castagliuolo I, Gino S, Caenazzo L (2020) Skin microbiome analysis for forensic human identification: what do we know so far? Microorganisms 8:873. https://doi.org/10.3390/microorganisms8060873

Tridico SR, Murray DC, Addison J, Kirkbride KP, Bunce M (2014) Metagenomic analyses of bacteria on human hairs: a qualitative assessment for applications in forensic science. Investig Genet 5:16. https://doi.org/10.1186/s13323-014-0016-5

Vallone C, Rigon G, Gulia C, Baffa A, Votino R, Morosetti G, Zaami S, Briganti V, Catania F, Gaffi M, Nucciotti R, Costantini FM, Piergentili R, Putignani L, Signore F (2018) Non-coding RNAs and endometrial cancer. Genes (Basel) 9(4):187. https://doi.org/10.3390/genes9040187

Virkler K, Lednev IK (2009) Analysis of body fluids for forensic purposes: from laboratory testing to non-destructive rapid confirmatory identification at a crime scene. Forensic Sci Int 188(1–3):1–17. https://doi.org/10.1016/j.forsciint.2009.02.013

Watanabe K, Akutsu T (2020) Evaluation of a co-extraction kit for mRNA, miRNA, and DNA methylation-based body fluid identification. Legal Med 42:101630. https://doi.org/10.1016/j.legalmed.2019.101630

Watanabe K, Taniguchi K, Toyomane K, Akutsu T (2022) A new approach for forensic analysis of saliva-containing body fluid mixtures based on SNPs and methylation patterns of nearby CpGs. Forensic Sci Int Genet 56:102624. https://doi.org/10.1002/elps.202000217.4

Weber M, Davies JJ, Wittig D, Oakeley EJ, Haase M, Lam WL (2005) Chromosome-wide and promoter-specific analyses identify sites of differential DNA methylation in normal and transformed human cells. Nat Genet 37(8):853–862. https://doi.org/10.1038/ng1598

Willis JR, Gabaldón T (2020) The human oral microbiome in health and disease: from sequences to ecosystems. Microorganisms 8(2):308. https://doi.org/10.3390/microorganisms8020308

Woerner AE, Novroski NMM, Wendt FR, Ambers A, Wiley R, Schmedes SE, Budowle B (2019) Forensic human identification with targeted microbiome markers using nearest neighbor classification. Forensic Sci Int Genet 38:130–139. https://doi.org/10.1016/j.fsigen.2018.10.003

Wojciuk B, Salabura A, Grygorcewicz B, Kędzierska K, Ciechanowski K, Dołęgowska B (2019) Urobiome: in sickness and in health. Microorganisms 7:548. https://doi.org/10.3390/microorganisms7110548

Xu H, Zhao Y, Liu Z, Zhu W, Zhou Y, Zhao Z (2012) Bisulfite genomic sequencing of DNA from dried blood spot microvolume samples. Forensic Sci Int Genet 6(3):306–309. https://doi.org/10.1016/j.fsigen.2011.06.007

Yang Y, Xie B, Yan J (2014) Application of next-generation sequencing technology in forensic science. Genomics Proteomics Bioinformatics 12(5):190–197. https://doi.org/10.1016/j.gpb.2014.09.001

Ye J, Xu M, Tian X, Cai S, Zeng S (2019) Research advances in the detection of miRNA. J Pharm Anal 9(4):217–226. https://doi.org/10.1016/j.jpha.2019.05.004

Zubakov D, Boersma AW, Choi Y, van Kuijk PF, Wiemer EA, Kayser M (2010) MicroRNA markers for forensic body fluid identification obtained from microarray screening and quantitative RT-PCR confirmation. Int J Legal Med 124(3):217–226. https://doi.org/10.1007/s00414-009-0402-3

Zubakov D, Kokmeijer I, Ralf A, Rajagopalan N, Calandro L, Wootton S, Langit R, Chang C, Lagace R, Kayser M (2015) Towards simultaneous individual and tissue identification: a proof-of-principle study on parallel sequencing of STRs, amelogenin, and mRNAs with the Ion Torrent PGM. Forensic Sci Int Genet 17:122–128. https://doi.org/10.1016/j.fsigen.2015.04.002

3 Technological Advancements in DNA Extraction and Quantification of Forensic Samples

3.1 Introduction

Biological samples found in a variety of biological fluids and tissues, including human blood, semen, saliva, epithelial cells, hair, bone, teeth, fingernails, and decomposed tissues, are included in the category of forensic biological evidence (Barbaro 2021). Non-human samples from plants, animals, microbes, and fungi may also need to be processed in addition to human samples. In most forensic biology practices, DNA extraction is a crucial step to generate a successful DNA profile, leading to the conviction of the suspect in the court. However, DNA extraction is a challenging step, profoundly so when handling samples in different circumstances that might cause degradation or inhibition to the DNA.

The DNA experts need to utilize the right technology to maximize the level of fitness and acceptance of the DNA profiles in the justice system. Although forensic DNA has advanced throughout the years, many subjects remain challenging to the DNA experts such as skeletal remains or samples contaminated by Chemical, Biological, or Radiological (CBR) agents, or processing of difficult samples like incomplete separation of spermic DNA from sperm-epithelial cell mixtures (Chong et al. 2021).

Touch DNA is an additional matter to address when it comes to low shredding or sweating individuals who has very low input of DNA in the crime scene. Basically, DNA extraction has three stages; (1) Lysis of cell membrane to release DNA into the lysis solution, (2) enzymatic digestion and/or denaturation of the proteins to unsequester DNA from the nucleoproteins, and (3) separation of DNA from other cellular components and inhibitors. So far, there is no single method that might be efficient for all different biological materials in variety of crime scene circumstances.

However, there are different literature reports in advanced technologies for numerous biological materials. Many facts need to be taken into account when following any procedure for forensic DNA lab: (1) fast DNA extraction method to support the workload, (2) effective method, (3) handling environmental contaminants, (4) lesser DNA degradation, and (5) reduced amount of DNA inhibitors that might affect the DNA amplification stage via polymerase chain reaction (PCR). DNA quantification assists in studying the efficiency of DNA extraction and the quality of DNA (Spitzer et al. 2022). It is the step where to further investigate the DNA evidence for downstream processing or report the results as negative case.

Prior to using any destructive analytical process, it is crucial to ascertain the complete amount of DNA that is available since it is critical in forensic analysis to preserve as much of the evidence as possible for retesting (Lin et al. 2018). Quantitative real-time PCR (qPCR) is one of the more recent approaches that can detect the presence of degraded DNA and PCR inhibitors, by

H. R. Dash et al., *Advancements in Forensic DNA Analysis*,
https://doi.org/10.1007/978-981-99-6195-5_3

comparing the results with the internal positive control (IPC) which can help identify potential causes of unfit genotyping outcomes and provide strategies for successful downstream typing (Lee et al. 2014). In general, it is highly advisable to have all the information about the DNA samples in our hands before proceeding for downstream applications. There are various types of available kits that detect the STRs as well as the YSTRs as discussed in Chapter 1. The usage of the kits depends on the case circumstances. In this chapter, many advanced technological practices in DNA extraction will be explained for a variety of samples. Also, the newest DNA quantification kits will be elucidated.

3.2 DNA Extraction

The genetic substance in humans and other organisms is called DNA. Nuclear DNA is the term for the majority of DNA such as that that is found in a cell's nucleus. Nuclear DNA is tightly packed in the chromosomes, and they normally contain 23 pairs of chromosomes (46 in total) in humans (Barbaro 2021). Normally, there is about 6 pg of DNA in every cell. DNA typing reveals the same DNA profile for diverse sources of DNA such as blood, sperm, skin cells, bones and teeth, and hair. This feature made DNA a reliable tool to use as forensic evidence in courts.

DNA extraction or isolation is defined as purifying DNA from all of the cell components other than DNA, for downstream processes. The first nucleic acid isolation was done in 1869 by the Swiss physician and biologist Johannes Friedrich Miescher (Dahm 2008). Following Miescher's devoted work, many experiments took place by other scientists to explore more about the nucleic acids, leading to the Watson and Crick discovery in 1953 and presentation of the three-dimensional double helix structure of the DNA molecule (Pray 2008). Overall, all these findings led to utilizing DNA as human identification tool besides fingerprinting analysis. The following factors affect the DNA extraction process: (1) the type of sample, (2) the wide range of substrates the sample may be placed on, (3) the amount of evidence being examined, (4) exposure to different environmental conditions, (5) variable degradation, (6) and the presence of PCR inhibitors (Lee and Shewale 2006).

Therefore, in order to achieve an efficient quantity of DNA samples free of inhibitors, it is imperative to utilize a procedure that can maximize DNA recovery from forensic evidence, ensuring the full integrity of DNA structure which can be used for downstream processing (Butler 2011). That is, the DNA extraction from hair sample is different than from blood sample. As a result, forensic evidence can include any combination of sample and substrate types in different circumstances. There are different characteristics of DNA extraction process that one must consider:

1. The system must allow acid nucleic extraction from a variety range of biological samples,
2. The system must allow maximum recovery of DNA in sufficient quantity and quality for downstream analysis,
3. The system must enable DNA isolation from trace samples such as skin cells or old samples,
4. The system must reduce the loss of available DNA,
5. The system must remove the inhibitors with PCR,
6. The system must not introduce any more contaminants,
7. The system must not alter nor degrade DNA sample,
8. The used system must be simple and rapid for daily basis,
9. The system must be amenable for automation and finally
10. The system must not use hazardous reagents which can be harmful for the DNA analysts (Barbaro 2021). There are number of sources of DNA samples which is given Table 3.1.

In this chapter, different extraction and purification platforms are explained that are used for daily basis work, i.e., Organic and the automated extraction such as AutoMate Express, EZ1/EZ2

Table 3.1 Sources of samples and the average amount of DNA obtained

Sample	Amount of DNA
Blood	30,000–40,000 ng/mL
Semen	150,000–300,000 ng/mL
Saliva	5000–12,000 ng/mL
Urine[a]	6–7000 ng/mL
Feces	8–17 ng/mg
Hair root	55–100 ng/root
Teeth[b]	15,000–20,000 ng/tooth
Bone	3–10 ng/mg
Tissue	50–500 ng/mg
Skin cells	Less than 100 pg/swab

[a] Male urine
[b] Molar teeth

Advanced XL, and Maxwell 16 Bench-top DNA extraction systems.

3.2.1 Organic DNA Extraction System

This process takes a lot of time and effort and it consists of five major steps:

- Breaking cells open to release the DNA.
- Separating DNA from proteins and other cellular debris.
- Precipitating the DNA with an alcohol.
- Cleaning the DNA.
- Confirming the presence and quality of the DNA.

Sodium dodecyl sulfate, Tris-Cl, and ethylene diamine tetraacetic acid (EDTA) are nonionic detergents that can be used to lyse cells. Centrifugation is then used to remove the cell debris. Proteins are subsequently denatured by protease treatment. Proteins are denaturated and precipitated from nucleic acid solution using organic solvents like chloroform, phenol, or a chloroform/phenol/isoamyl alcohol mixture (the ratio of these three solvents is 25:24:1). Denatured proteins are then eliminated using centrifugation and washing procedures. Unwanted RNA is removed with RNAse treatment. For the purpose of concentrating DNA, ice-cold ethanol is precipitated (Gupta 2019).

Additionally, DNA in the samples can be further purified and concentrated by recovering it from the aqueous phase using a centrifugal filter unit (such as the Vivacon or Amicon devices) (Comey et al. 1994). When monovalent cations are present in a modest quantity, nucleic acid precipitate forms (salt). Centrifugation can be used to recover the precipitate, which is then redissolved in Tris-EDTA (TE) buffer or double-distilled water (Gupta 2019). Figure 3.1 illustrates the steps of organic extraction. Other techniques include salting out, anion exchange technology, silica-based technology, magnetic separation, and cesium chloride density gradients. DNA attaches reversibly to magnetic beads that have been coated with DNA-binding antibodies (Gupta 2019).

In order to have an assessment of DNA quality and yield, spectrophotometry or gel electrophoresis are used to evaluate the quality and yield of DNA. Spectrophotometry measures the amount of light absorbed by the sample at particular wavelengths in order to estimate the DNA concentration. Nucleic acids have an absorption peak at 260 nm. For dsDNA, the A260/A280 ratio is 1.8. Protein contamination is indicated by a ratio of less than 1.7.

Although this approach is still one of the most effective and dependable, it takes a long time, involves hazardous chemicals, and increases the risk of contamination and sample handling

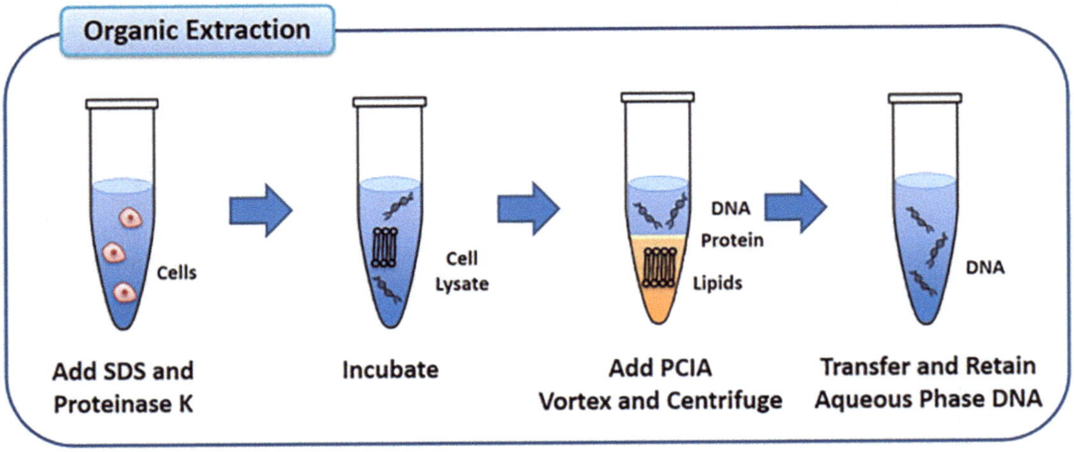

Fig. 3.1 Organic DNA extraction procedure using phenol:chloroform:isoamyl alcohol. (Adapted from McKiernan and Danielson 2017)

because it requires more hands on and frequent tube transfers (Köchl et al. 2005).

3.2.2 Automated DNA Extraction System

The extraction system is intended to operate commercial kits that extract proteins, RNA, and DNA from a variety of sources, such as:

- Soil
- Plants
- Bacteria
- Blood
- Urine
- Saliva
- Feces
- Skin
- Human remains
- Hair
- Bones and teeth
- Sweat.

The system can be set up to use any one or all of your preferred separation techniques, such as:

- vacuum filtration
- pressure filtration
- magnetic bead separation
- centrifuging.

3.2.2.1 EZ1&2 DNA Investigator Kit, Qiagen

The EZ1&2 DNA Investigator Kit reproducibly automates purification of genomic DNA from reference and casework samples in human identity testing (HID) in EZ1 Advanced XL or EZ2 Connect Fx Instruments (Qiagen, Hilden Germany). The kit can be used with the EZ2 Connect Fx (1–24 samples), EZ1 Advanced XL (1–14 samples per run), EZ1 Advanced (1–6 samples per run), and BioRobot EZ1 (1–6 samples per run) instruments. Purification is fast and efficient. Purified DNA performs well in downstream analyses with high signal-to-noise ratios, such as quantitative PCR and STR analysis.

Optimized chemistries for automated DNA purification with higher yields are offered by the EZ1&2 DNA Investigator Kit. Higher peaks in the STR analysis indicate more sensitive detection. High-quality DNA which is provided by magnetic beads method is suitable for direct use in numerous downstream applications including quantitative PCR, STR analysis, and next-generation sequencing (NGS). Different samples can be used to process using the EZ1&2 DNA

Investigator Kit such as bones and teeth, hair, blood, soil, sperm, skin, saliva, fired and unfired ammunition, etc. (Scherer et al. 2022).

3.2.2.2 Maxwell FSC DNA IQ Casework Kit, Promega

With the help of DNA IQ chemistry and the Maxwell FSC Instrument (Promega, Wisconsin, USA), casework samples including blood stains, semen stains, hair, cigarette butts, tissues, and trace DNA samples can be extracted in a consistent, automated manner. Automated DNA extraction reduces labor expenses and saves lab personnel time so they may focus on casework analysis.

A plastic cartridge and newly designed plunger are used in the Maxwell FSC DNA IQ Casework Kit to enable DNA elution in a final volume of no more than 50 μL. The pre-filled cartridge comes with DNA IQ Lysis Buffer, Resin, and Wash Buffer, and the kit also comes with DNA IQ Elution Buffer to ensure that the DNA is stored properly. The Maxwell FSC Instrument and Maxwell RSC 48 Instrument, both of which come with a surface tablet and an easy-to-use interface, are compatible with the Maxwell FSC DNA IQ Casework Kit. One to 16 samples can be processed by the Maxwell FSC Instrument, and one to 48 samples can be processed by the Maxwell RSC 48 in a single run.

3.2.2.3 PrepFiler Express Forensic DNA Extraction Kit, Thermo Fisher Scientific

The AutoMate Express Forensic DNA Extraction System (Thermo Fisher Scientific, USA) is compatible with this kit which processes 1–13 samples in a single run. The vast majority of common sample types used in forensic laboratories, including bodily fluids on various substrates containing FTA paper, cotton swabs, cotton cloth, denim, and many others, are suitable for it.

In order to increase the quantity and quality of DNA extracted from forensic materials and thus increase the possibility of obtaining the most data from subsequent Short-Tandem Repeat (STR) analysis, the PrepFiler chemical was specifically designed. The kit is created to increase the yield and general purity of DNA obtained from both routine and challenging forensic samples. It delivers high-quality DNA that is free of PCR inhibitors and suited for further uses including STR analysis and quantitative real-time PCR. The kit uses the special PrepFiler LySep Column, the substrate and lysate can be separated easily and quickly and with pre-filled cartridges that are ready to use and make setup quick and simple.

It has a closed system that reduces the possibility of contamination and transposition errors. The kit has rapid implementation with pre-validated, pre-programmed instrument protocols and integrated solution with innovative reagents, consumables, and automation to provide time-saving, cost-effective, and reliable DNA extraction. Furthermore, it improves downstream genotyping (STR) success rate to ensure profile quality.

3.3 Comparison Between Different Extraction Platforms

The AutoMate Express (Thermo Fisher Scientific, USA), a device frequently used in forensic laboratories, was used to compare the performance of the EZ2 Connect Fx (Qiagen, Hilden Germany). Both equipment processed duplicates of a variety of mock forensic casework materials, and the outcomes of DNA purification and yield recovery were compared. Different samples were used such as hair, saliva, blood, touch samples, and inhibited samples. All samples for the EZ2 Connect Fx were processed using the Investigator Lyse&Spin Basket Kit for substrate removal and lysis. All samples were processed using a PrepFiler LySep Column for the AutoMate Express (Duchamp et al. 2022a).

Following manufacturer's instructions, DNA purification of the samples was done using the EZ1&2 DNA Investigator Kit on the EZ2 Connect Fx and the PrepFiler Express Forensic DNA Extraction Kit on the AutoMate Express. For all of the examined saliva and blood dilutions, the concentration of DNA recovered from blood and saliva samples using the EZ2 Connect Fx method is higher than that obtained using the AutoMate Express. However, variations in the results of Automate Express and EZ2 Connect Fx are shown in hair roots, touch DNA and cigarette butts (Duchamp et al. 2022a).

Another validation study was done in comparison between the EZ2 Connect Fx (Qiagen, Hilden Germany) and Maxwell FSC Instrument (Promega, Wisconsin USA) using mock forensic samples of blood, saliva, touch samples, inhibited samples, and cigarette butts. The study reveals that the concentration of DNA recovered from swabs with the EZ2 Connect Fx procedure is higher than with the Maxwell FSC for all the saliva, blood dilutions tested as well as all the inhibited samples and touch samples. However, there are slight variations in results of the samples from the fingerprints on glass slides and cigarette butts (Duchamp et al. 2022b). More studies need to be conducted using more challenging samples such as ancient DNA from bone and teeth samples as well as human remains.

3.4 Quality Control of the Extracted DNA

To ensure the quality of DNA extracted, numerous quality control methods are utilized. The concentration and purity of a DNA sample can be determined using the widely used technique of spectroscopic analysis. Spectrophotometry is the absorbance of a sample is measured at several wavelengths, commonly at 260 and 280 nm. The purity of the DNA sample is assessed using the relationship between the absorbance at 260 and 280 nm. Gel electrophoresis is a method for examining and contrasting DNA sample size and integrity. An electric field is used to move the DNA through an agarose gel after it has been put onto the gel. Ethidium bromide, which intercalates into the DNA and fluoresces under UV light, can be used to see how the DNA is migrating. Fluorometry is a technique for calculating the concentration of nucleic acids by monitoring the sample's fluorescence after being activated by a particular wavelength of light. Fluorometry makes use of dyes with strong fluorescence intensities that precisely bind to nucleic acids. The Qubit Fluorometer is a device that measures the quantity of DNA and RNA in a sample by using fluorescent dyes. The concentration of DNA samples can be determined using this simple and accurate procedure. The Bioanalyzer is a device that separates and examines DNA, RNA, and protein materials using electrophoresis. It can offer thorough details regarding the size, consistency, and purity of a DNA sample.

3.5 DNA Quantification

For accurate DNA amplification and STR profiling, it is crucial to quantify the DNA in forensic samples. Numerous techniques have been created to quantify DNA, including fundamental UV spectrometry, gel-based techniques, dye labeling, blotting, and, most recently, DNA amplification techniques (polymerase chain reaction, PCR). Early methods only measured total DNA, while more recent methods can test only human DNA, omitting non-human DNA (foodstuff, animal, or bacterial contamination). These more recent assays are perfect for the busy forensic laboratory because they can be quicker and less expensive than conventional techniques. The authors of a study evaluated both established and emerging quantification approaches, and they also offer recent methodologies they have created based on PCR of Alu sequences (Nicklas and Buel 2003).

For a number of reasons, it is crucial to determine the quantity and quality of human DNA retrieved from biological evidence. Firstly, the quality (purity and length) and quantity of the extracted DNA might vary significantly depending on the source and extraction technique. This has an impact on the downstream technique since the amount of input DNA and its relative length can dictate which genotyping method—mini-STR typing, standard STR) typing, or mitochondrial DNA sequencing—should be used. Second, before using any destructive analytical technique, it is crucial to calculate the total amount of DNA accessible because it is crucial in forensic analysis to retain as much of the evidence as possible for retesting. Last but not least, the outcomes of the preliminary quantitative and qualitative analyses provide a more knowledgeable interpretation of the outcomes.

Real-time PCR is one of the more recent quantitative approaches that can detect the presence of damaged DNA and PCR inhibitors, which can help identify potential causes of weak genotyping outcomes and provide strategies for successful downstream typing. Quantitative real-time PCR (qPCR), since it was first introduced to the forensic field a decade ago, has played an integral role in the workflow of DNA analysis (Lee et al. 2014). Not only can qPCR determine the total amount of amplifiable human DNA present in a forensic casework sample, but also the degree of inhibition present with its Internal Positive Control (IPC). In general, the easier it is to understand and analyze the sample, which increases the possibility of successful DNA typing, the more information there is accessible. The improvement of analysis precision and the expansion of result information content have been the two fundamental objectives throughout the history of the development of quantitative methods (Lee et al. 2014).

In this chapter, different qPCR kits will be discussed namely Investigator Quantiplex HYres Kit from Qiagen, Investigator Quantiplex Pro from Qiagen, Quantifiler Trio DNA Quantification Kit from Thermo Fisher Scientific (previously Applied Biosystems), and PowerQuant System from Promega.

The earlier generations of these qPCR kits, such as the Quantifiler Human and Y Human Male DNA Quantification Kits evaluate the total quantity of amplifiable human or male DNA in a sample and quantify a single target in the genome, respectively (Green et al. 2005). The next-generation kits, such as the PowerQuant System, Quantifiler Trio DNA Quantification Kit, Investigator Quantiplex HYres Kit and Investigator Quantiplex Pro, are able to quantify autosomal and male DNA simultaneously, providing additional details on the proportions of male to female human DNA in a sample (Frégeau and Laurin 2015; Ip and Lai 2016).

To quantify the degree of DNA degradation in a DNA sample, the so-called third generation kits Quantifiler Trio and PowerQuant even included a longer autosomal or degradation amplicon (D), in their quantification systems in addition to the traditional autosomal DNA amplicon (Auto) (Ewing et al. 2016; Holt et al. 2016). Longer (D) targets on the template DNA sample are more likely to degrade during PCR amplification than are shorter (Auto) targets because of their bigger sizes.

As a result, the autosomal and degradation amplicons' ratio (Auto)/(D), often referred to as the degradation index or "DI" below, can be used to determine the level of degradation in a DNA sample (Lin et al. 2018) (Table 3.2).

In a comparison study conducted in 2018 between Investigator Quantiplex HYres Kit, Quantifiler Trio DNA Quantification Kit and PowerQuant System (Lin et al. 2018), the results revealed that the most accurate kit was PowerQuant. Also, better correlations and lesser variances between autosomal and male DNA quantification were provided by Quantifiler Trio DNA Quantification Kit and Powerquant systems (Lin et al. 2018). The findings showed that Quantifiler Trio, PowerQuant, and Quantiplex HYres all perform better than Quantifiler Human and Y (Thermo Fisher Scientific, USA). In addition to having greater sensitivity and consistency, these three new generation kits also provide simultaneous detection of both autosomal and male DNA in a faster turnaround time. The amplicon for big autosomal or degradation is even included in Quantifiler Trio and PowerQuant, giving an estimate of the level of degradation and inhibition in a sample at the stage of quantification (Lin et al. 2018).

In another comparison study conducted in 2018 between Investigator Quantiplex Pro Kit, Quantifiler Trio DNA Quantification Kit, PowerQuant System and InnoQuant HY (InnoGenomics Technologies), the results declared that for high-quality DNA samples with low inhibitor concentrations, all of the study's kits functioned equally. The Investigator Quantiplex Pro Kit was most tolerant of all spiked samples, with the exception of those with high

Table 3.2 Details of the four commonly used DNA quantification kits

Quantification kit	Amplicon/target	Size (bp)	Reaction time (min)
PowerQuant	Autosomal	84	~60
	Degradation	294	
	Male	81 & 136	
	IPC	435	
Quantifiler Trio	Autosomal (small)	80	~60
	Autosomal (large)	214	
	Male	75	
	IPC	130	
Quantiplex HYres	Autosomal	146	~60
	Male	129	
	IPC	200	
Quantiplex Pro	Autosomal (small)	91	~60
	Autosomal (large)	353	
	Male	81	
	IPC	434	

salt concentrations, and provided more precise quantification findings in samples with greater amounts of inhibitors (Sorensen Holmes et al. 2018).

3.6 DNA Quantification Instrumentation

There are two optical technologies commonly used to quantify nucleic acids: UV-Vis measurement and fluorescence measurement. The photometric measurement of nucleic acids is based on the intrinsic absorptivity properties of nucleic acids (DNA and RNA). When an absorption spectrum is measured, nucleic acids absorb light with a characteristic peak at 260 nm. The fluorometric measurement of nucleic acids is based upon the use of fluorogenic dyes that bind selectively to DNA or RNA.

However, for more accurate interpretation of Human DNA quantification, qPCR is used. There are several qPCR instruments that are used for forensics and human identification testing (HID) such as QuantStudio 3 and QuantStudio 5 Real-Time PCR Systems (Thermo Fisher Scientific, USA), Rotor-GeneQ (Qiagen, Hilden Germany) and Real-Time PCR System (Thermo Fisher Scientific, USA).

There are no current comparison studies that show which system works the best. However, some system have more features that make it more preferable than others. In order to evaluate the performance of a real-time PCR reaction, the following parameters must be evaluated:

- **Dynamic range**
  To accurately determine the efficiency of a PCR reaction, a 5-log dilution series must be performed. A slope of $-3.3 \pm 10\%$ reflects an efficiency of $100 \pm 10\%$. A PCR reaction with lower efficiency will have lower sensitivity.
- **R2 value**
  R^2, a statistical term that denotes how well one result can predict another, is another crucial factor in assessing PCR efficiency. The value of the threshold cycle (C_t) can be utilized to precisely anticipate the value of x when R^2 is 1 (Fig. 3.2a). The value of x cannot be predicted from the value of y if R^2 is 0 (Fig. 3.2b). A R^2 score of >0.99 indicates a strong correlation between two values.
- **Precision**
  The most widely used unit of precision is the standard deviation, which is the square root of variance. The standard deviation is smaller if numerous data points are close to the mean and

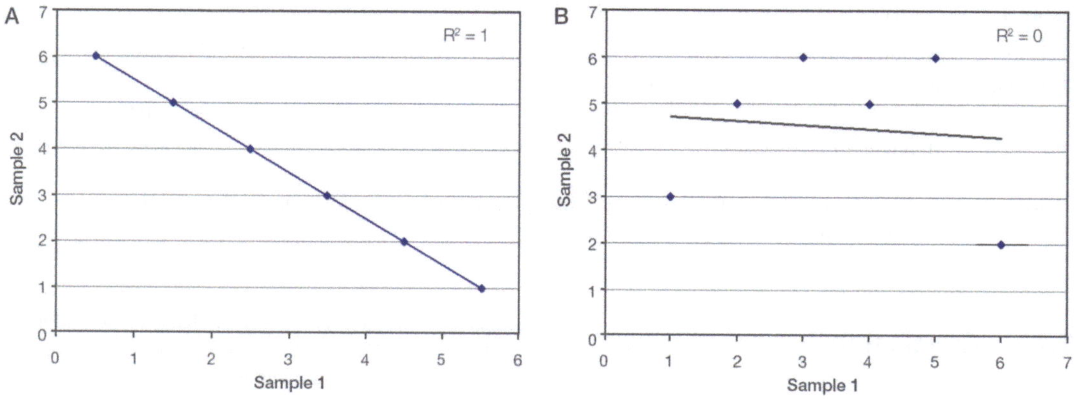

Fig. 3.2 Examples of R^2 values calculated for two straight lines. (**a**) There is a direct relationship between x and y values. (**b**) There is no linear relation between x and y values. (Adapted from Thermo Fisher Scientific)

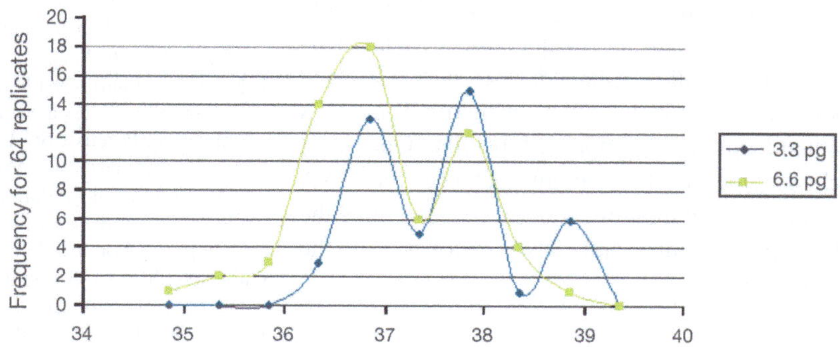

Fig. 3.3 Poisson distribution for low copy number. (Adapted from Thermo Fisher Scientific)

larger if many data points are far from the mean.

- **Sensitivity**

  The highest level of sensitivity has been attained by any system, regardless of C_t value, that is capable of amplifying and detecting one copy of the beginning template. The fact that a normal distribution of templates is not anticipated is another crucial factor to take into account when detecting extremely low copy numbers. Instead, a Poisson distribution is used (Fig. 3.3), which predicts that in many replicates with an average of one copy of the starting template, 37% should actually have none, 37% should have just one copy, and 18% should have two copies.

  Therefore, a large number of replicates are required to give statistical significance and get beyond the Poisson distribution barrier for reliable low copy detection. When comparing various reaction conditions, the performance of a PCR reaction is assessed using efficiency, R^2, accuracy, and sensitivity. In Table 3.3, all factors are listed for an optimum evaluation of qPCR.

  There are some adapted protocols to skip the DNA extraction and purification method and continue directly to qPCR. One of the is the cutting-edge sample collection tool is PE-Swab (Liu 2014). A PE-Swab can be used to create a tiny sample punch that can be used in a qPCR reaction for quantification. Accurate DNA quantification can be acquired from a sample without the requirement for DNA extraction and purification after tweaking the punch size and the quantification software baseline setting. There are other methods that terminate the need to DNA extraction and quantification and goes directly to PCR amplification.

Table 3.3 Performance evaluation of real-time PCR

Factors	Recommendations	Criteria
Efficiency	Serial dilution with 5-log dilutions	Slope: ~ −3.3
		$R^2 > 0.99$
Precision	Minimum of 3 replicates	Standard deviation <0.167
Sensitivity	High replicate number of reactions for low copy number sample input due to Poisson distribution	Statistical test analysis

3.7 Conclusion

In this chapter, different DNA extraction methods were discussed including the latest technology of magnetic beads for DNA purification. Comparison validation studies were done to illustrate which instruments were given more DNA yield than others. Also, DNA quantification methods were thoroughly explained, with new advanced kits and instruments such as using qPCR in DNA quantification for improved results. One must choose the most optimum methods for DNA extraction and quantification in order to have the best performance and DNA recovery of the evidence.

References

Barbaro A (2021) Challenges in DNA extraction from forensic samples. In: Dash HR, Shrivastava P, Lorente JA (eds) Handbook of DNA profiling. Springer, Singapore, pp 1–20

Butler JM (2011) Advanced topics in forensic DNA typing: methodology. Academic Press, Oxford

Chong KWY, Thong Z, Syn CKC (2021) Recent trends and developments in forensic DNA extraction. Wiley Interdiscip Rev Forensic Sci 3(2):e1395

Comey CT, Koons BW, Presley KW, Smerick JB, Sobieralski CA, Stanley DM, Baechtel F (1994) DNA extraction strategies for amplified fragment length polymorphism analysis. J Forensic Sci 39(5): 1254–1269

Dahm R (2008) Discovering DNA: Friedrich Miescher and the early years of nucleic acid research. Hum Genet 122(6):565–581

Duchamp V, Haley J, Liberty A, Krsicka O, Faethe S, Nastainczyk-Wulf M (2022a) Comparison of two DNA extraction platforms for use in forensic casework applications: EZ2® Connect Fx versus AutoMate Express™. www.qiagen.com

Duchamp V, Haley J, Liberty A, Krsicka O, Golob N, Dierig L (2022b) Comparison of two DNA extraction platforms for use in forensic casework applications: EZ2® Connect Fx versus Maxwell® FSC Instrument. www.qiagen.com

Ewing MM, Thompson JM, McLaren RS, Purpero VM, Thomas KJ, Dobrowski PA, DeGroot GA, Romsos EL, Storts DR (2016) Human DNA quantification and sample quality assessment: developmental validation of the PowerQuant® system. Forensic Sci Int Genet 23:166–177

Frégeau CJ, Laurin N (2015) The Qiagen Investigator® Quantiplex HYres as an alternative kit for DNA quantification. Forensic Sci Int Genet 16:148–162

Green RL, Roinestad IC, Boland C, Hennessy LK (2005) Developmental validation of the Quantifiler™ real-time PCR kits for the quantification of human nuclear DNA samples. J Forensic Sci 50(4):809–825

Gupta N (2019) DNA extraction and polymerase chain reaction. J Cytol 36(2):116–117

Holt A, Wootton SC, Mulero JJ, Brzoska PM, Langit E, Green RL (2016) Developmental validation of the Quantifiler® HP and Trio Kits for human DNA quantification in forensic samples. Forensic Sci Int Genet 21: 145–157

Ip SC, Lai K-M (2016) An evaluation of the precision of QIAGEN quantiplex HYres kit as a screening tool on low level touch DNA evidence. Eur J Forensic Sci 3(2):61–67

Köchl S, Niederstätter H, Parson W (2005) DNA extraction and quantitation of forensic samples using the phenol-chloroform method and real-time PCR. Methods Mol Biol 297:13–30

Lee SB, Shewale JG (2006) DNA extraction methods in forensic analysis. In: Encyclopedia of analytical chemistry: applications, theory and instrumentation. Wiley, Chichester, pp 1–18

Lee SB, McCord B, Buel E (2014) Advances in forensic DNA quantification: a review. Electrophoresis 35(21–22):3044–3052

Lin S-W, Li C, Ip SC (2018) A performance study on three qPCR quantification kits and their compatibilities with the 6-dye DNA profiling systems. Forensic Sci Int Genet 33:72–83

Liu JY (2014) Direct qPCR quantification of unprocessed forensic casework samples. Forensic Sci Int Genet 11: 96–104

McKiernan H, Danielson P (2017) Molecular diagnostic applications in forensic science. In: Patrinos GP (ed) Molecular diagnostics. Elsevier, London, pp 371–394

Nicklas JA, Buel E (2003) Quantification of DNA in forensic samples. Anal Bioanal Chem 376(8): 1160–1167

Pray L (2008) Discovery of DNA structure and function: Watson and Crick. Nat Educ 1(1):100

Scherer M, Terramagra D, Liberty A, Prochnow A (2022) Developmental validation of the EZ1&2™ DNA Investigator® Kit on the EZ2® Connect Fx. Forensic Sci Int Genet Suppl Ser 8:350–351

Sorensen Holmes A, Houston R, Elwick K, Gangitano D, Hughes-Stamm S (2018) Comparison of four commercial qPCR kits for analyzing inhibited and degraded forensic samples. https://www.qiagen.com/

Spitzer A, Einot N, Voskoboinik L, Balas S, Amiel M, Oz C (2022) Choosing a forensic quantification kit to optimize DNA profiling workflow. JFI 72(2):142–156

4 Advanced Emerging Techniques for Forensic DNA Analysis: STRs, SNPs, and mtDNA Analysis

4.1 Introduction

In the year 2022, the prestigious Nobel prize in Physiology or Medicine was given to Svante Pääbo for studying our extinct ancestors' DNA. Svante Pääbo did what was once thought impossible: He figured out how to extract DNA from the ancient bones. He successfully extracted DNA from a fossilized 40,000 years old finger bone and carried out subsequent analyses. He found that gene transfer had occurred from these now extinct hominins to *Homo sapiens* (Brown et al. 2022). This ancient flow of genes to the present-day humans has many physiological relevance today, including affecting how our immune system reacts to infections. The discovery and methodologies used by Svante Pääbo is an inspiration to the forensic DNA analysts in the laboratory as they lament not able to generate a successful DNA profile from a relatively recent forensic biological sample.

A new word has been added to the previously existing three mottos of the Olympic motto and now it reads as *Citius*, *Altius*, and *Fortius-Communis*, i.e., faster, higher, stronger- together. Over the last few decades, the DNA technologies as well as the forensic DNA technologies have emerged at a rapid pace. Most of the newly developed techniques are envisioned to meet these characteristics of the Olympic motto. The forensic DNA fraternity is in the search of any such techniques which is robust, shows rapid results and includes interdisciplinary techniques such as STR-SNPs, DIP-STRs, and many advanced techniques for better resolution of discrimination (Castella et al. 2013; Wei et al. 2018). With the advent of Next Generation Sequencing (NGS) technology, forensic DNA analysis have been further revolutionized. NGS technology has allowed the sequencing of whole mitochondrial genome in a much lesser time. Thus, revealing a better resolution in discriminating between two individuals in comparison to the sequence data of hypervariable regions of the mitochondrial DNA (Yuan et al. 2020). In this regard, this chapter discusses the advanced technologies of STRs, SNPs and mtDNA sequencing for forensic DNA analysis.

4.2 STR Techniques and Its Advancements

Exploitation of human genetic variation has been for the key for forensic DNA analysis. Until the 1980s, the identification of the nature of samples followed by individualization was highly dependent on conventional approaches such as histology, microscopy, immunology, biochemistry, and serology. Useful of conventional genetic markers for the purpose of individualization included blood group antigens, red cell isozymes, serum proteins, variants of hemoglobin, and human leukocyte antigen (HLA) system. The degree of variability of these protein markers

H. R. Dash et al., *Advancements in Forensic DNA Analysis*,
https://doi.org/10.1007/978-981-99-6195-5_4

was very limited and it was immensely difficult to achieve a sufficiently high level of discrimination. Due to their low level of resolution, such techniques are considered powerful tools for exclusion of a possible source. However, the discovery of hypervariable polymorphism in DNA sequences, called variable number tandem repeats (VNTRs) or minisatellites by Sir Alec Jeffreys, revolutionized the field of forensic DNA analysis.

VNTRs are the sites of a DNA fragment with size differences due to variations in the number of repeats within the human population. Such size differences in a VNTR locus can be determined using the Restriction Fragment Length Polymorphism (RFLP) technique. This technique was considered to be the gold standard for individualization and identification purposes at that time. Both single locus and multi-locus RFLP analysis can be performed. In this technique, DNA was isolated followed by fragmentation of DNA using restriction enzymes. Each individual restriction enzyme cuts at its own specific sequences whenever found along the DNA chain. Furthermore, gel electrophoresis is performed where the DNA fragments are separated by their size. After separation of DNA fragments, they are blotted from the gel onto a nylon membrane employing the technique called as Southern Blotting. Single or multi-locus probes are made radioactive and many copies of the probe(s) are used to combine with a specific DNA sequence on the nylon membrane. Furthermore, X-ray film is placed over the membrane to detect and image the radioactive probe pattern, the technique of which is called as autoradiography. The overall technique of RFLP exploring VNTR markers is shown in Fig. 4.1.

Very soon after its inception, the RFLP technique was found to be less sensitive, technically laborious, and require high quantity of good quality DNA. In forensic casework, it is immensely difficult to find a huge quantity of good quality DNA. In this connection the discovery of Polymerase Chain Reaction (PCR) technique in 1983 came as a rescue, and the technique was termed as amplified length polymorphism (AFLP). Besides other disadvantages, the VNTR markers are not highly polymorphic (Teama 2018). The number of possible alleles of a VNTR marker is limited in human beings. Single locus probes can't discriminate between individuals. Use of multi-locus probes is a better alternative. However, there is a limitation in including the number of markers in a single reaction using RFLP technique. Even after using multi-locus probes, sufficient level of discrimination is not achieved (Jeffreys et al. 1991).

4.2.1 Characteristics and Advantages of STR Markers

The above shortcomings of VNTR markers and the inception of PCR technique has led to the search for alternative markers for superior use in identification and individualization especially in the forensic biological samples. In this regard, the use of microsatellites or Short Tandem Repeat (STR) markers came into the limelight and such markers were successfully used in the identification of a murder victim from 8-year-old skeletal remains (Hagelberg et al. 1991). In this case, the skeletal remains of a 15-year-old Caucasian female murdered in 1981 was discovered in 1989. Though 5 μg DNA was extracted from the bone samples, 90% of these DNA were found to be exogenous in origin, due to bacterial and fungal contamination. Besides, the extracted DNA was severely degraded and the fragments were less than 300 nucleotides long. RFLP using VNTRs did not yield a DNA profile in this case. Even the amplification of minisatellites using PCR technique was not possible. Hence for the first time, PCR amplification of the six polymorphic microsatellite markers was carried out in the forensic DNA sample and the amplified fragments were separated by agarose gel electrophoresis. All the six STR markers were amplified in the sample and the DNA profiles generated from the putative mother and father were compared for identification (Hagelberg et al. 1991). After this successful application of PCR-based STR technique, it is till now considered to be the gold standard markers for identification and

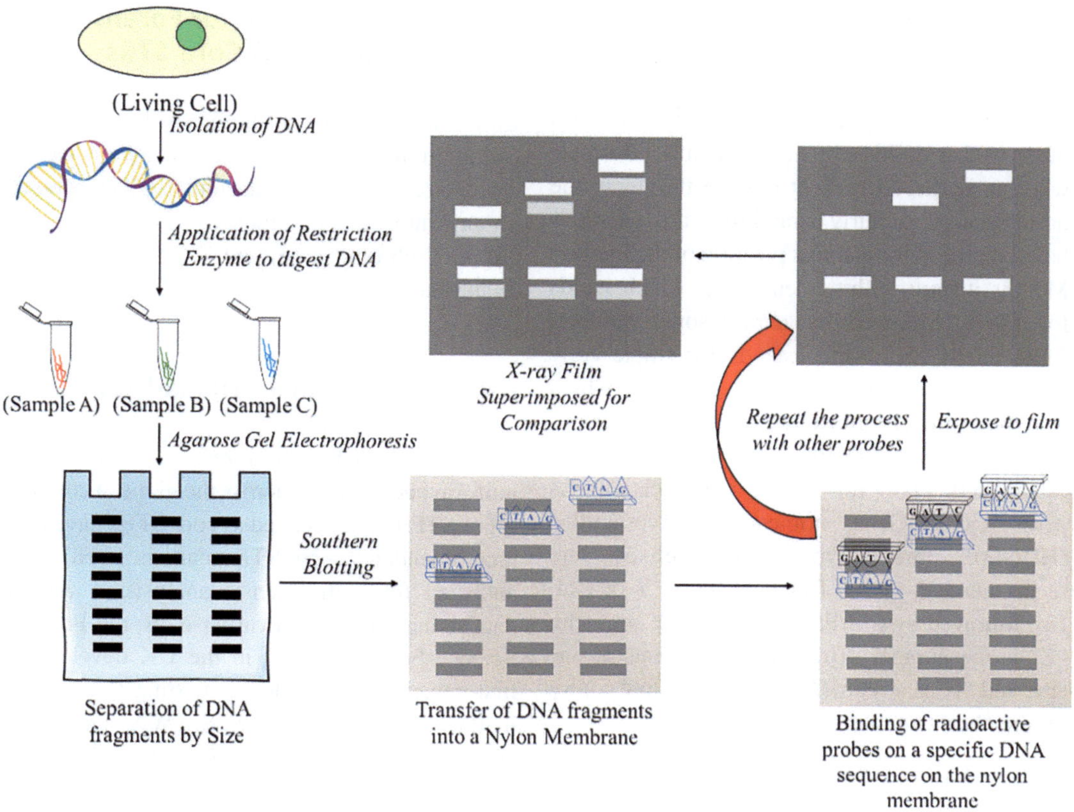

Fig. 4.1 Detailed schematic presentation of RFLP technique for forensic DNA analysis

individualization in the field of forensic DNA analysis.

STRs also called as simple sequence repeats (SSRs) or microsatellites feature accordion-like stretches of DNA. Each contains core repeat units of between two and seven nucleotides in length. These core repeat units are tandemly repeated from approximately a half dozen to several dozen times in human populations (Butler 2007). One STR occurs per 2000 bp in the human genome and there exists more than one million of STR markers in the human genome (Fan and Chu 2007). Abundance of STR markers in the human genome and its compatibility with PCR make it the method of choice among the forensic scientists. As a PCR-enabled method, ~0.5 ng of DNA sample is required to carry-out the requisite amplifications. Besides, relatively shorter PCR products are formed of approximately 100–500 bp length for the STR markers. Such small amplicon sizes are useful for testing forensic samples containing degraded DNA. Use of fluorescent dyes in conjugation with PCR primers allows simultaneous amplification of different STR markers and their detection by different amplicon sizes and their corresponding fluorescent dyes. Use of multiple STR loci in a single reaction increases the discrimination power of a DNA profile without consuming a high quantity of DNA (Stanley et al. 2020).

4.2.2 Nomenclature of STR Markers

A uniform pattern of nomenclature exists for the STR markers commonly used in forensic DNA analysis. If a marker is present inside a gene, the name of the STR marker is as per the name of the gene. For example, STR present in first intron of tyrosine hydroxylase gene is named as TH01.

Similarly, CSF1PO (sixth intron of proto-oncogene c-fms), FGA (third intron of alpha fibrinogen gene), vWA (40th intron of von Willebrand factor gene), and TPOX (tenth intron of thyroid peroxidase gene) are named. Besides, when a marker is present outside the gene, the nomenclature is fairly straightforward which is based on the chromosomal position of the marker. Mostly, it starts with the letter "D," which stands for DNA, followed by chromosome number, followed by "S," which stands for single copy sequence, followed by the number which designates the chromosomal position. For example, D5S818, DYS391, and DXS8237 are established STRs for human identification. As per HUGO Gene nomenclature committee (HUGO), three types of prefixes are recommended, i.e., D followed by X, Y or Chr. No. followed by S/Z/F. "S' is a unique segment, "Z" is the number of repetitive segments at one site, and "F" is a segment that has multiple copies on the chromosome.

4.3 Types of STR Markers

The length of the repeat unit determines the type of a STR marker. Di-, tri-, tetra-, and penta-nucleotide STR markers have two, three, four, and five nucleotides as their core repeat unit, respectively. Besides the length of the repeat unit, STR markers also vary in terms of their confirmation. Some of the repeat unit of STR markers contain repeat units of identical length and sequence. These STRs are called as simple STRs. Similarly, some of the STR markers contain two or more adjacent simple repeats as their repeat units; they are called as Compound STRs. When several repeat units of variable repeat unit length and variable intervening sequences are present, they are called complex STRs. Though it is generally considered that the number of repeat units of an STR locus varies in different individuals, many times the alleles contain an incomplete repeat unit sequence; such a condition is called a micro-variant (Dash et al. 2021a). The most common example of a microvariant allele is 9.3 of TH01.

4.4 STR Loci Used for Forensic DNA Analysis and Core STRs

Based on their chromosomal position and application in forensic DNA analysis, they can be categorized as either Autosomal STRs (present on any autosomal chromosomes), Y-STRs (present on Y-chromosome) and X-STRs (present on X-chromosome).

4.4.1 Autosomal STR Markers

In order to make the forensic DNA analysis uniform and effective across the globe, a common set of STR markers called as core STRs have been recommended. Such STR markers need to be analyzed for both forensic analysis as well as databasing purpose. In earlier days, the Forensic Science Services (FSS) in the UK developed a first-generation multiplex consisting of four STR markers, i.e., TH01, FES/FPS, vWA, and F13A1 (Kimpton et al. 1994). This tetraplex STR markers had a matching probability of 1 in 10,000 approximately. Soon it was replaced with a hexaplex consisting of TH01, vWA, FGA, D8S1179, D18S51, D21S11 and a sex determining marker. This hexaplex was called as Second-Generation Multiplex (SGM) and showed a matching probability of 1 in 50 million (Kimpton et al. 1996). The gender identification marker recommended in this SGM was Amelogenin.

Soon the United States started its efforts to create a National DNA Database known as the Combined DNA Index System (CODIS). In a meeting of the STR project during 13–14 November, 1997 a total of 13 STR markers were selected for their inclusion in the CODIS National DNA database. The selected 13 STR markers includes CSF1PO, FGA, TH01, TPOX, vWA, D3S1358, D5S818, D7S820, D8S1179, D13S317, D16S539, D18S51, and D21S11. Later in 1999 a European Standard Set (ESS) of STR loci were selected. This original ESS included FGA, TH01, vWA, D3S1358. D8S1179, D18S51, and D21S11. This set contains the hexaplex recommended by FSS and an additional marker, i.e., D3S1358. Interpol embraced the core STRs recommended by ESS and renamed them as Interpol Standard Set of Loci (ISSOL). Later in April, 2005, five new STR markers were included in the ESS by European Network of Forensic Science Institutes (ENFSI) which included D2S441, D10S1248, D22S1045, D1S1656, and D12S391 (Gill et al. 2006). Thus, a total tally of 12 STR markers for European databases was made which was accepted by the European Union in November, 2009.

In a parallel manner, as the US National Database crossed nine million profiles by 2010, the laboratory of Federal Bureau of Investigation (FBI) explored the possibility to expand its core set of STR loci. After a huge brainstorming, the Expanded CODIS markers were recommended and came into practice since 1st January, 2017. In addition to the 13 earlier CODIS markers, seven additional markers, i.e., D1S1656, D2S441, D2S1338, D10S1248, D12S391, D19S433, D22S1045 were included to constitute the expanded CODIS (Hares 2015). The major reasons of expanding the core CODIS loci from 13 to 20 are (1) to reduce the likelihood of adventitious matches, (2) to increase international compatibility, and (3) to increase discrimination power to aid missing persons cases. The list of STR markers included in the expanded CODIS and their corresponding characteristics is given in Table 4.1.

4.4.1.1 Parameters Used to Assess the Usability of an STR Marker

Various population specific parameters of the STR markers should be analyzed prior to their application in the forensic DNA analysis. These parameters can be divided into two types: the paternity parameters and the forensic parameters. Both the types of parameters should be considered for the assessment of the usability of a STR markers in casework and databasing purpose. Some of these parameters are discussed below.

4.4.1.1.1 Allele Number and Total Possible Genotypes

The number of possible genotypes of a genetic marker reflects its usefulness in forensic applications. The number of possible genotypes of a marker is determined by the number of possible alleles. Genotypes are of two types, i.e., homozygous and heterozygous. If the number of possible alleles of a genetic marker is "n," then the number of possible homozygous genotypes is also "n." Whereas, the number of possible heterozygous genotypes is $n(n - 1)/2$. The total number of possible genotypes becomes the sum of number of possible homozygous genotypes and the number of possible heterozygous genotypes. Hence, higher the number of alleles observed in a genetic marker, higher is its usefulness in forensic DNA analysis in terms of higher number of possible genotypes (Dash et al. 2021b).

4.4.1.1.2 Match Probability

The probability of matching (PM) can be defined as the probability that two randomly selected individuals will have identical genotype. It is calculated individually for each STR marker tested and the cumulative match probability of all the STR markers tested can be calculated to find the match probability of a DNA profile. The lower the match probability of a STR marker, the more useful this marker is for identification purpose. The match probability can be calculated using the following formula:

$$p_{\mathrm{M}} = \sum_{k=1}^{m} p_k^2$$

where p_{M} is the match probability, p_k represents the frequency of each distinct genotype, and m is the number of distinctive genotypes. The combined random match probability of all the loci

Table 4.1 List of STR markers and their characteristics included in expanded CODIS

Sl. No.	Locus	Chromosomal location	Physical position	GenBank accession	Type	Repeat motif
1	D1S1656	1q42	Chr 1230.905 Mb	G07820	Compound	TAGA
2	TPOX	2p25.3	Chr 2 1.493 Mb	M68651	Simple	AATG
3	D2S441	2p14	Chr 2 68.239 Mb	AC079112	Compound	TCTA/TCAA
4	D2S1338	2q35	Chr 2218.879 Mb	AC010136	Compound	TGCC/TTCC
5	D3S1358	3p21.31	Chr 3 45.582 Mb	AC099539	Compound	TCTA/TCTG
6	FGA	4q31.3	Chr 4155.509 Mb	M64982	Compound	CTTT/TTCC
7	D5S818	5q23.2	Chr 5123.111 Mb	AC008512	Simple	AGAT
8	CSF1PO	5q33.1	Chr 5149.455 Mb	X14720	Simple	AGAT
9	D7S820	7q21.11	Chr 7 83.789 Mb	AC004848	Simple	GATA
10	D8S1179	8q24.13	Chr 8125.907 Mb	AF216671	Compound	TCTA/TCTG
11	D10S1248	10q26.3	Chr 10131.093 Mb	AL391869	Simple	GGAA
12	TH01	11p15.5	Chr 11 2.192 Mb	D00269	Simple	TCAT
13	vWA	12p13.31	Chr 12 6.093 Mb	M25858	Compound	TCTA/TCTG
14	D12S391	12p13.2	Chr 12 12.450 Mb	G08921	Compound	AGAT/ AGAC
15	D13S317	13q31.1	Chr 13 82.692 Mb	AL353628	Simple	TATC
16	D16S539	16q24.1	Chr 16 86.386 Mb	AC024591	Simple	GATA
17	D18S51	18q21.33	Chr 18 60.949 Mb	AP001534	Simple	AGAA
18	D19S433	19q12	Chr 19 30.416 Mb	AC008507	Compound	AAGG/ TAGG
19	D21S11	21q21.1	Chr 21 20.554 Mb	AP000433	Complex	TCTA/TCTG
20	D22S1045	22q12.3	Chr 22 37.536 Mb	AL022314	Simple	ATT

tested is the product of the individual value of each loci tested.

4.4.1.1.3 Power of Discrimination

Power of discrimination is the probability that any two randomly selected individuals in a population can have different genotypes. A STR marker having a higher value of power of discrimination is considered to be more useful for forensic identification purpose in that population. The formula for calculating power of discrimination is as follows:

$$\text{For a single STR locus}: p_{\mathrm{D}} = 1 - p_{\mathrm{M}}$$

$$\text{For multiple loci combined}: P_{\mathrm{Dcomb}} = 1 - \prod_{i=1}^{n} (1 - P_{\mathrm{D}i})$$

where PoD is the power of discrimination of a single locus, p_{M} is the match probability of that locus, PoD_i is the power of discrimination of the individual locus and Π sign stands for multiplication.

4.4.1.1.4 Power of Exclusion

Power of exclusion (PE) defines the fraction of the individuals that is different from a randomly selected individual. It signifies the power of a STR marker to exclude a person to be the biological father. The calculated value of PE differs in each case. Thus, the average value may be calculated to find the power of exclusion of a single locus. The power of exclusion can be calculated using the following formula:

$$\text{For a single locus}: \mathrm{PE} = h^2\left(1 - 2hH^2\right)$$

where h is the heterozygosity and H is the homozygosity at the locus.

$$\text{For several loci}: \mathrm{PE}_{\mathrm{comb}} = 1 - \prod_{l=1}^{L} (1 - \mathrm{PE}_l)$$

where L is the number of the loci, PE_l is the exclusion probability for the lth locus and the Π sign represents multiplication.

4.4.1.1.5 Polymorphic Information Content

Polymorphic information content (PIC) indicates the polymorphic level of a STR locus. PIC can be calculated using the following formula:

$$\mathrm{PIC} = 1 - \sum_{i=1}^{n} p_i^2 - \left(\sum_{i=1}^{n} p_i^2\right)^2 + \sum_{i=1}^{n} p_i^4$$

where P_i is the frequency of each distinct allele and n is the number of distinct alleles.

4.4.1.1.6 Typical Paternity Index

Paternity index is the calculated value for a single STR locus that determines the statistical weight of that locus in favor or against parentage in a case. The higher the PI in a STR marker, the more useful it becomes in that population for paternity testing. The paternity index (PI) of a STR marker can be calculated using the following formula:

$$\mathrm{PI} = (H + h)/2H$$

where H is the number of homozygotes and h is the number of heterozygotes.

4.4.1.1.7 Genetic Diversity

The genetic diversity (GD) calculation provides a measurement to demonstrate that a genetic marker is showing variation in the population. A STR marker with high genetic diversity value is considered to be more useful for identification of an individual in that population. The genetic diversity of a STR marker can be calculated using the following formula:

$$\mathrm{GD} = \frac{n\left(1 - \sum p_i^2\right)}{n - 1}$$

where n is the sample size, p_i is the allele frequency of the ith allele (King et al. 2022).

4.4.1.1.8 Heterozygosity

Heterozygosity reflects the marker specific genetic variation in the studied population. High heterozygosity infers lots of genetic variability in that marker; whereas, low heterozygosity shows little genetic variability. Mostly the observed level of heterozygosity (Ho) is compared with the expected level of heterozygosity (He) under Hardy–Weinberg equilibrium (HWE). If Ho is lower than He, we assume the existence of micro-evolutionary forces such as inbreeding. In contrary, if Ho becomes higher than He, an isolate breaking effect can be suspected such as the mixing of two previously isolated populations. Expected heterozygosity is also called as Gene Diversity.

4.4.2 Y-Chromosome STRs

STRs found on Y-chromosome were first used in solving a case of sexual assault in 1990 in Germany. In this case, a woman was murdered after having sexual intercourse and blood sample and vaginal swab were collected from the deceased, besides collecting blood sample from the suspect. A differential extraction process was carried out to separate sperm fraction from the epithelial fraction and the extracted DNA was subjected to conventional hybridization techniques using DNA probes. As this technique failed, an attempt was made with PCR-based amplification of four STRs, two located on chromosome 12, one on chromosome 6 and one on Y-chromosome. Y-STR was amplified in the sperm fraction and showed a different length of amplicon in comparison to the amplicon generated from the suspect's blood. As a result of unequivocal result of Y-STR, the suspect was exonerated and released from the jail (Roewer 2019).

4.4.2.1 Applications of Y-STR Analysis

Since the solving of the above-mentioned case, many STR markers have also been discovered on Y-chromosome for their wide use in forensic DNA analysis. In comparison to the biparental inheritance of autosomes, the Y-chromosome shows uniparental inheritance (Fig. 4.2). A majority of the Y-chromosome is inherited from the father to the son without recombination. Thus, random mutation becomes the only source of genetic variation between paternal relatives. This unique nature of Y-chromosome and its STRs have allowed the use of Y-STRs in forensic DNA analysis when autosomal analysis generates

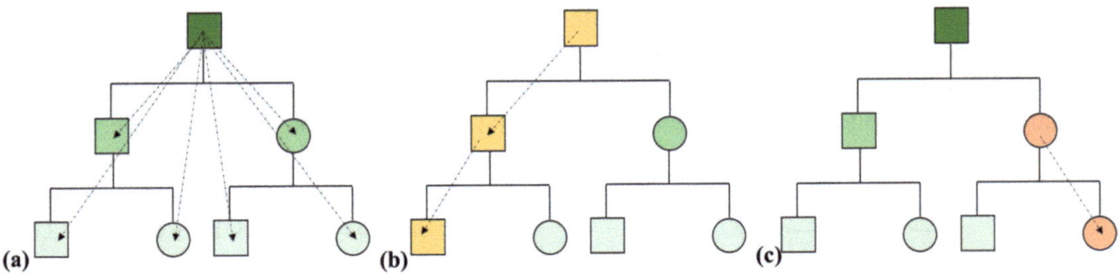

Fig. 4.2 Inheritance pattern of various genetic markers (**a**) biparental inheritance of recombining autosomal markers, (**b**) uniparental inheritance (father to son) of Y-chromosome markers, (**c**) uniparental inheritance (mother to child) of mtDNA

limited results. Such an application scenario may exist as, (1) When high level of female DNA is present in comparison to male DNA as happens in sexual assault cases and mixture samples, (2) Determination of sex in AMELY deletion cases, (3) To exclude male suspects from involvement in crime, (4) Identification of paternal lineage of male perpetrators, (5) Highlighting multiple male contributors to a trace, and (6) To provide investigative leads for finding unknown male perpetrators.

4.4.2.1.1 Analysis of DNA with High M: F Ratio

In many cases, mostly in sexual assault cases, forensic scientists encounter mixed body fluids stains. In sexual assault cases, most of the samples are collected from the victim's body by the swabbing method. Thus, the chance of getting epithelial cells of the victim increases. The problem becomes worse if the victim bleeds during penetration or she is under menstrual period at the time of incidence. With an increase in the time gap between the date of incidence and date of sampling, the samples collected from victim's body contains fewer perpetrator's cells in comparison to the female cells. All these factors cumulatively affect the male (M) to female (F) ratio of the extracted DNA from the samples. In a sample with high ratio of M:F DNA, autosomal STR analysis may not yield an interpretable mixed autosomal STR profile because during the PCR process the high amount female DNA is preferentially amplified in comparison to the low quantity male DNA which goes undetected in the capillary-electrophoresis process (Star et al. 2016). However, in such samples Y-STR analysis plays a useful role as such markers are found exclusively on Y-chromosome and a DNA profile from the perpetrator is generated.

Advancement of DNA quantification techniques in the form of quantitative real-time PCR (RT-PCR) has revolutionized the field of forensic DNA analysis. It analyzes the extracted DNA samples to give useful information such as total human DNA content, degradation of DNA, and presence or absence of PCR inhibitors in the samples. In addition to that, analysis of extracted DNA by RT-PCR can provide the ratio of male-to-female sources, which becomes highly useful for the forensic scientists for choosing the suitable methodology for further downstream processing.

4.4.2.1.2 AMELY Deletion Cases

The amelogenin gene produces heterogeneous mRNAs due to alternate splicing, subsequently generating heterogenous group of amelogenin proteins responsible for tooth enamel matrix formation. Two homologues of amelogenin, AMELX and AMELY, exhibit length polymorphisms. In comparison to AMELY (+AAAGTG), a 6 bp deletion is found in AMELX (−AAAGTG) thus distinguishing between male and female DNA sources (Dash et al. 2020). Since the discovery of the amelogenin marker, most of the commercially available kits employ this marker for sex determination in a forensic sample (Sullivan et al. 1993). Although amelogenin has been considered a well-established marker for forensic sex determination purposes, the efficacy of the amelogenin marker came into question when cases of false detection of males as females started surfacing. The amelogenin marker may give false negative results due to allele dropouts resulting from either deletions or mutations in the gene region or mutation in the primer binding site. The amelogenin failure cases and their studies have increased several fold in the last decade (Thangaraj et al. 2002).

In such a scenario, most of the commercially available kits have started using some markers present on Y-chromosome as alternate sex determining markers. In AMELY deletion conditions where a male sample generates only an X peak in the DNA profile, these additional sex determining markers present on Y-chromosome are amplified and may determine the correct sex of the sample source. To address the AMELY deletion conditions, most of the newest kits have started using either DYS391, DYS570 or DYS576 as additional STR markers for the appropriate determination of the sex of the individual. A few of the kits have started using a Y-InDel marker besides the Y-STR markers. The small size of the Y-InDel amplicon (81–86 nt) provides an added advantage of this marker for its amplification in degraded samples besides its usefulness as an alternate sex determining marker (Ludeman et al. 2018).

4.4.2.1.3 Exclusion of Male Suspects

Due to its non-recombination phenomenon and uniparental inheritance nature, the Y-STR profile cannot individualize a person. The X- and Y-markers are lineage markers. Y-STR DNA profile is the same in case of two patrilineal relatives. Some unrelated males can also have the same Y-STR DNA profile (Liu et al. 2016). Although it is not possible to statistically include a suspect as a contributor to the Y-STR DNA profile generated from the questioned sample, it plays a pivotal role in excluding the suspect. When the Y-STR DNA profile generated from the questioned sample does not match with the reference sample profile of the suspect, the suspect can be easily excluded for its involvement. Thus, in compromised samples (as described in the previous section), where autosomal analysis provides limited information, Y-STR analysis can perform well in excluding a suspect by looking for the mismatches in those Y-STR DNA profiles.

4.4.2.1.4 Identification of Paternal Lineage

As the surname of an individual indicates the family, tribe or community that person belongs to, the Y-STR profile can also be useful for determination of ancestry between individuals. Ninety-five percent of the Y-chromosome is the non-recombining region with the X-chromosome. It is well-preserved through many generations in the paternal lineage. With the use of new generation kits having 27 or more Y-STR markers, a high haplotype diversity has been observed (Purps et al. 2014). The occurrence of such a huge haplotype diversity allows forensic DNA examiners to exclude male perpetrators from being involved in a crime through non-matching of Y-STR haplotypes. Besides, Y-typing also allows the analyst to identify the paternal lineage by matching Y-STR haplotypes. It allows the analyst to gather investigative leads which will be useful for the investigating agencies.

4.4.2.1.5 Cases Involving Multiple Male Contributors

Interpretation of a mixed profile is always a challenge for forensic DNA examiners. The situation becomes more complicated in cases involving more than one perpetrator - mostly in gang rape cases. Generation of multiple peaks at an autosomal STR marker complicates the attribution of each peak to a specific suspect. In such cases, Y-STR analysis can be highly useful to discriminate among the suspects as their Y-STR haplotypes may differ. Besides, in sexual assault cases, the female victim does not contribute to the Y-STR profile, thus reducing the total number of genotypes to the mixed profile. In this regard, a recent study on many sexual assault cases showed that one tenth of such cases would have remained inconclusive without the application of Y-STR haplotyping. This study also reported that Y-STR haplotyping is three time more appropriate to identify multiple male contributors than using autosomal STR analysis alone (Purps et al. 2015).

4.4.2.1.6 Investigative Leads for Unknown Male Perpetrators

In certain countries like the Netherlands, the DNA profiles are kept after standard DNA profiling for their potential investigative purposes. If a familial link cannot be established using database searching of autosomal STR profiles, searching criteria may be modified to patrilineal societies using Y-STR haplotype data. This might be helpful in establishing the possible patrilineal relatives of an offender whose profile is not available in the database (Kayser 2017).

4.4.2.2 Interpretation of Y-STR Profile

Y-STR loci present in the non-recombining Y (NRY) region of the Y-chromosome are inherited as a complete block. Mendel's Law of independent assortment does not apply to these STR markers as they are present in linkage disequilibrium with each other. Thus, unlike autosomal STRs, the product rule of probability cannot be applied to determine the frequency of multi-locus Y-STRs using the frequency of individual STR alleles. Rather, a simple counting method is commonly used to estimate the frequency of a particular Y-STR haplotype. The frequency of a haplotype is the number of times the haplotype is observed in the database divided by the total number of samples present in the database. Other methods have also been established to estimate the frequency of occurrence of a particular Y-STR haplotype. One such method is based on the mismatch distribution approach which allows the evaluation of the proportion of pairs of Y-STR haplotypes that are prone to become identical by state (IBS), in one generation, by recurrent mutation (Pereira et al. 2002).

However, there exists a severe limitation with the availability of sufficient number of sample data in the database for appropriate calculation of a Y-STR haplotype. When the existing database is of limited size, a confidence interval is applied to the haplotype frequency. However, a haplotype frequency can also be reported without associated confidence intervals stating a factual statement regarding the observation limited to the particular database only. In this case, the upper 95% confidence limit can be calculated using the formula:

$$p + 1.96\sqrt{p(1-p)/n}$$

where $p = x/n$ (n is the database size and x is the number of times the haplotype is observed in the database).

If a haplotype is not available in the database and observed for the first time, the following formula can be used to calculate the 95% confidence limit:

$$1-(0.05)^{1/n}$$

where n is the size of the database.

Most of the studies have confirmed the non-existence of any linkage between the autosomal STRs and Y-STRs (Budowle et al. 2005). Hence when both autosomal and Y-STRs are analyses are performed together, the autosomal STR profile frequency can be multiplied with the upper confidence limit of the Y-STR haplotype frequency to estimate the overall frequency of the combined profiles.

4.4.2.3 RM-YSTRs

The most common limitation of using Y-STR analysis is its inability to distinguish between close or distant patrilineal relatives of the suspects as the contributor of the biological material. In a case involving close or distantly related patrilineal relatives, Y-STR analysis is of limited help due to the low mutation rates of these Y-STR markers which is only a few mutations per 1000 generations per locus (Goedbloed et al. 2009). Such a problem arises in all the cases with inclusion with Y-STR profiles. In a recent study, when 186 relatively newly discovered Y-STR markers were analyzed, 13 of them showed substantially higher mutation rate than that of the average mutation rates. These markers were designated as rapidly mutating (RM) Y-STRs. These 13 Y-STRs also showed their ability to improve

Table 4.2 Characteristics of 13 rapidly mutating (RM) Y-STR markers (Ballantyne et al. 2012)

Marker	Repeat type	Repeat motif	Mutation rate
DYF387S1	Tetra, Complex	(AAAG)3(GTAG)1(GAAG)4N16(GAAG)9 (AAAG)13	1.58×10^{-2}
DYS399S1	Tetra, Complex	(GAAA)3N7–8(GAAA)10–23	7.73×10^{-2}
DYS403S1a	Tetra, Complex	A: (TTCT)10–17N2–3(TTCT)3–17	3.10×10^{-2}
DYS403S1b	Tetra, Complex	B: (TTCT) 12N2 (TTCT) 8 (TTCC) 9 (TTCT) 14 N2 (TTCT)3	1.18×10^{-2}
DYF404S1	Tetra, Complex	(TTTC)10–20N42(TTTC)3	1.24×10^{-2}
DYS449	Tetra, Complex	(TTCT)13–19N22(TTCT)3N12(TTCT)13–19	1.22×10^{-2}
DYS518	Tetra, Complex	(AAAG)3(GAAG)1(AAAG)14–22(GGAG)1 (AAAG)4N6(AAAG)11–19N27(AAGG)4	1.84×10^{-2}
DYS526a	Tetra, Complex	(CCCT)3N20(CTTT)11–17(CCTT)6–10N113 (CCTT)10–17	2.72×10^{-3}
DYS526b	Tetra, Complex	(CCCT)3N20(CTTT)11–17(CCTT)6–10N113 (CCTT)10–17	1.25×10^{-2}
DYS547	Tetra, Complex	(CCTT)9–13T(CTTC)4–5N56(TTTC)10–22N10 (CCTT)4(TCTC)1 (TTTC)9–16N14(TTTC)3	2.36×10^{-2}
DYS570	Tetra, Simple	(TTTC)14–24	1.24×10^{-2}
DYS576	Tetra, Simple	(AAAG)13–22	1.43×10^{-2}
DYS612	Tri, Complex	(CCT)5(CTT)1(TCT)4(CCT)1(TCT)19–31	1.45×10^{-2}
DYS626	Tetra, Complex	(GAAA)14–23N24(GAAA)3N6 (GAAA)5 (AAA)1 (GAAA)2–3(GAAG)1(GAAA)3	1.22×10^{-2}
DYS627	Tetra, Complex	(AGAA)3N16(AGAG)3(AAAG)12–24N81 (AAGG)3	1.23×10^{-2}

the paternal lineage resolution in world-wide populations. The list and characteristics of these 13 RM-YSTRs is described in Table 4.2.

4.4.3 X-STRs

The X-chromosome (Chr-X) shows a unique pattern of inheritance in comparison to other genetic markers. It has both uniparental and biparental inheritance. Normal male individuals possess a single X-chromosome, which is inherited from the mother, whereas, in case of normal females, out of two X-chromosomes, one is inherited from each of the parents. Such a unique nature and pattern of inheritance of X-chromosome has been proved useful for its analysis in forensic cases. X-chromosome STR analysis is highly useful in kinship cases with deficient parent samples. Similarly, in a case of a relationship establishment between a father and a daughter, X-STR analysis may be highly useful as 100% transmission of the father's X-chromosome takes place to the daughter. Chr-X testing is highly useful in missing persons identification and disaster victims identification cases mostly in sample limiting conditions. In parent-offspring relationships that involves at least one female, X-STR analysis is deemed to be highly useful. The other useful applications of X-STR analysis are illustrated in Fig. 4.3. In a case report, it has been shown that, both the mother and the female child of a father-daughter incest showed consistent X-STR profiles (Dash et al. 2023). Hence, X-STR

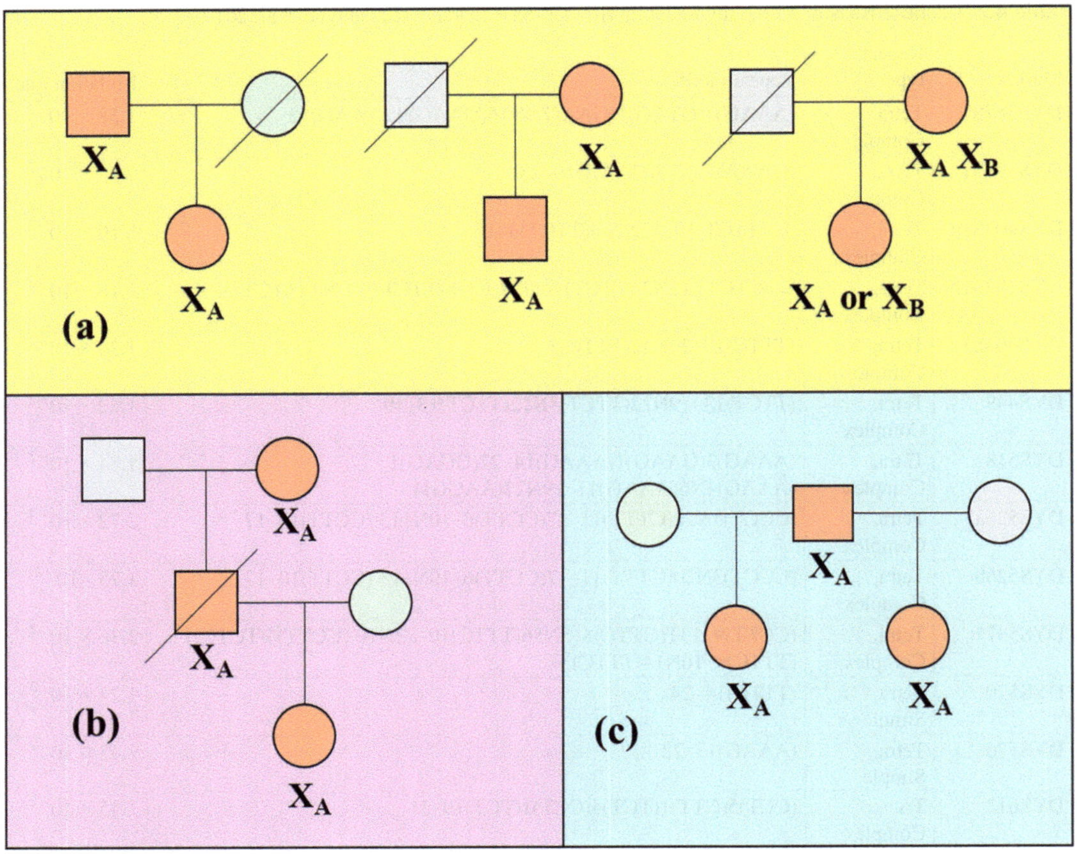

Fig. 4.3 Usefulness of X-STR in forensic DNA analysis (**a**) sample limiting conditions, (**b**) paternal grandmother-granddaughter, and (**c**) half-sisters testing

analysis also has its wide applications in complicated kinship scenarios such as incest.

4.4.3.1 Commercially Available New Generation Kits

4.4.3.1.1 Autosomal STR Kits

To ease the application of forensic DNA analysis, many commercial kits are now available. After the expansion of the CODIS core loci to 20, many new generation kits are available from different manufacturers. Most of these kits work on 6-dye chemistry. These commercially available kits can be assessed in terms of their robustness, specificity, rapidity, sensitivity, precision, reproducibility and repeatability, case-type samples, concordance, stutter, DNA mixtures, and PCR-based procedures. Some of the new generation autosomal STR analysis kits include GlobalFiler®, VeriFiler™ Plus, Investigator® 24Plex QS, PowerPlex® Fusion 6C, and SureID® PanGlobal human DNA identification kit. Most of these new generation kits accommodate the 20 expanded CODIS loci, Extended European Standard Set of loci (ESSL), miniSTR loci and alternate sex determining markers. GlobalFiler® kit multiplexes 24 STR markers with an amplification time of around 80 min. A validation study of this kit showed that a full profile was consistently obtained with an input DNA of 125 pg or more (Ludeman et al. 2018). The VeriFiler™ Plus kit simultaneously amplifies 23 autosomal STR loci, a quality indicator system, and two sex determining markers. A higher sensitivity of 62.5 pg input or more has been reported for the VeriFiler™ Plus kit (Green et al. 2021). The Investigator® 24Plex QS is a 6-dye multiplex system which coamplifies 22 polymorphic STR markers, amelogenin as the sex determining marker, DYS391 as the alternative sex determining marker, and the Quality Sensors, QS1 and QS2. The validation study of this kit shows a consistent amplification of 100% markers with 125 pg or more DNA input (Kraemer et al. 2017). The PowerPlex® Fusion 6C System is a 27-locus, 6-dye, multiplex containing the expanded core CODIS loci, and two rapidly mutating Y-STR markers. On both the 3500 and 3130 Genetic Analyzers, 100% alleles were reported to be called at as low as 250 pg of DNA (Ensenberger et al. 2016). The SureID PanGlobal system is a 27-locus multiplex system containing 24 autosomal STR loci, two Y-chromosome markers and amelogenin as the sex determining marker. A sensitivity of 125 pg of template DNA has been reported in the validation study of PanGlobal system (Liu et al. 2017). A comparative account of STR markers present in different new generation kits is given in Table 4.3.

4.4.3.1.2 Y-Chromosome STR Kits

The limitations of using Y-STRs over autosomal STRs is profound. Low levels of discrimination of Y-STR markers have limited the use of these markers as lineage markers. With the discovery of rapidly mutating (RM) Y-STRs, most of the commercially available new generation Y-STR kits have increased the number of Y-STRs besides including the RT-YSTRs. Such kits are deemed to have a wider application in forensic DNA analysis in terms of their usefulness in sexual assault cases, paternal lineage analysis as well as individualization of the individual. In this regard, Y Filer plus (Thermo Scientific) allows the simultaneous amplification of 27 Y-STRs including three highly polymorphic Y-STR loci (DYS460, DYS481, and DYS533) and seven rapidly mutating Y-STR loci (DYF387S1a/b, DYS449, DYS518, DYS570, DYS576, DYS627). This kit has a sensitivity of 125 pg of input DNA and improved peak balance can be observed with a male DNA versus female DNA mixture ratio of >1:1000 (Gopinath et al. 2016). PowerPlex Y-23 system (Promega Corp.) simultaneously amplifies 23 Y-STR markers including six highly discriminating Y-STR loci (DYS481, DYS533, DYS549, DYS570, DYS576, and DYS643) and two RM-YSTRs. The sensitivity of this kit has been reported to be 62.5 pg of input DNA. Complete Y-STR profiles have been detected in the mixed samples with 62.5 pg of male DNA in a background of 400 ng of female DNA (Thompson et al. 2013). Investigator Argus Y-28 QS system (Qiagen) amplifies the established 27 Y-STR markers and a quality sensor for better interpretation of the data. This kit also includes six rapidly mutating Y-STRs for better discrimination. A kit sensitivity of 100 pg of male DNA both in the presence and absence of female DNA has been reported (Vraneš et al. 2022). The SureID® PathFinder Plus is a new 6-dye, 41-plex Y-STR system which includes 14 rapidly mutating Y-STR loci (DYS449, DYS481, DYS518, DYS527a/b, DYS533, DYS549, DYS570, DYS576, DYS627, DYF387S1a/b, and DYF404S1), and 10 low-medium mutation loci (DYS388, DYS444, DYS447, DYS460, DYS522, DYS557, DYS593, DYS596, DYS643, and DYS645). The kit sensitivity has been reported to be 250 pg and a female cross reactivity has not been observed in a mixture of 1 ng male DNA with 1 μg of female DNA (Fan et al. 2021).

4.4.3.1.3 X-Chromosome STR Kits

Though X-STRs can be explored in many sample limiting conditions, they are not widely used in routine forensic DNA laboratories in comparison to autosomal and Y-STR kits. Fewer kits are available in the market for the analysis of X-STRs. Some of them include Investigator Argus X-12 QS Kit (Qiagen, Valencia, CA), SFX19™ PCR Amplification kit (Sorenson Forensics), AGCU X19 STR reagents (AGCU ScienTech Inc., Wuxi, Jiangsu, China), and 17 X-STR multiplex system. Investigator Argus X-12 QS Kit allows co-amplification of 12 X-STRs arranged in four linkage groups, i.e., 1: Xp22 (DXS8378, DXS10135, DXS10148), 2: Xp11 (DXS7132, DXS10074, DXS10079), 3:

Table 4.3 A comparative account of the genetic markers available in the new generation STR multiplex kits

Genetic Markers	GlobalFiler®	VeriFiler™ Plus	Investigator® 24Plex QS	PowerPlex® Fusion 6C	SureID® PanGlobal
D13S317	■	■	■	■	■
D7S820	■	■	■	■	■
D5S818	■	■	■	■	■
CSF1PO	■	■	■	■	■
D1S1656	■	■	■	■	■
D12S391	■	■	■	■	■
D2S441	■	■	■	■	■
D10S1248	■	■	■	■	■
D18S51	■	■	■	■	■
FGA	■	■	■	■	■
D21S11	■	■	■	■	■
D8S1179	■	■	■	■	■
vWA	■	■	■	■	■
D16S539	■	■	■	■	■
TH01	■	■	■	■	■
D3S1358	■	■	■	■	■
D2S1338	■	■	■	■	■
D19S433	■	■	■	■	■
TPOX	■	■	■	■	■
D22S1045	■	■	■	■	■
SE33	■		■	■	■
D6S1043		■			■
Penta D		■		■	■
Penta E		■		■	■
Amelogenin	■	■	■	■	■
DYS391	■		■	■	■
DYS570				■	
DYS576				■	
Y_{indel}	■	■			■
IQCS		■			
IQCL		■			
QS1			■		
QS2			■		

Xp26 (HPRTB, DXS10101, DXS10103), and 4: Xp28 (DXS10134, DXS7423, DXS10146). Besides the gender determining marker amelogenin, the autosomal marker D21S11 and an internal PCR control QS1 are also present (Scherer et al. 2015). SFX19™ PCR Amplification kit simultaneously amplifies 19 X-STR loci including DXS8378, DXS10074, DXS7423, DXS10148, DXS10159, DXS10134, DXS7424, DXS10164, DXS10162, DXS7132, DXS10079, DXS6789, DXS101, DXS10103, DXS10101, HPRTB, DXS6809, DXS10075, and DXS10135. AGCU X19 STR reagents probe 19 X-STRs found in seven clusters namely, cluster 1 (DXS10148, DXS10135, DXS8378), cluster 2 (DXS10159, DXS10162, DXS10164, DXS7132), cluster 3 (DXS10079, DXS10074, DXS10075), cluster 4 (DXS6809, DXS6789), cluster 5 (DXS7424, DXS101), cluster 6 (DXS10103, HPRTB, DXS10101), and cluster 7 (DXS10134, DXS7423) (Zhang et al. 2016). Another kit, the 17 X-STR panel has also been validated comprising the X-STRs DXS8378, DXS9898, DXS7133, GATA31E08, GATA172D05, DXS7423, DXS6809, DXS7132, DXS9902, DXS6789, DXS10079,

DXS6801, DXS6799, DXS6800, DXS10075, DXS6807, and DXS6803 (Prieto-Fernandez et al. 2016).

4.5 Mitochondria DNA Analysis and Its Advancements

Every nuclear cell contains two copies of genomic DNA (paternal and maternal), whereas 100s of copies of mitochondrial DNA (mtDNA) are present in each cell. Because of its high copy number, most of the forensic biological samples with degraded genomic DNA may contain mitochondrial DNA which can provide useful information upon its analysis. Forensic biological samples such as bone, teeth, and hair may yield a very low quantity of DNA where conventional autosomal STR analysis cannot be carried out. In such samples mitochondrial DNA analysis plays an important role. As illustrated in Fig. 4.2, mtDNA is useful as a lineage marker which is transferred in the maternal lineage.

4.5.1 Structural Characteristics of mtDNA

Mitochondria are thought to have originated from incorporated α-purple bacteria. It is a small circular genome found in the mitochondria. Mitochondria, being the energy house of the cell, vary in quantity from 100s to 1000s as per the physiological state of the cell. The average number of mitochondria per cell is estimated to be in the 500s (Satoh and Kuroiwa 1991). The size of mtDNA is approximately 16,569 base pairs weighing 10^7 Daltons. It is a histone-free circular double-stranded DNA molecule. The total number of nucleotides of mtDNA vary due to the presence of small insertions or deletions. A dinucleotide repeat at positions 514–524 is (AC)5 in most individuals, whereas, it may vary from (AC) 3 to (AC)7 (Szibor et al. 1997). In comparison to the single copy of the 6.4 billion bp (3.2 billion bp from each parent) of nuclear DNA, mtDNA makes up about 0.25% of the total DNA content of a cell. The difference between nuclear DNA and mtDNA is highlighted in Table 4.4.

Mitochondria genome codes for 37 gene products useful in oxidative phosphorylation. These gene sequences are found in the coding region of the mitochondria genome which includes 13 proteins, 2 ribosomal RNAs (rRNA), and 22 transfer RNAs (tRNA). The genes are very tightly/economically packaged without the presence of any introns or a few non-coding nucleotides are present within the coding regions. Besides, a 1122 bp "control" region is also present which contains the origin of replication for mtDNA replication. This "control" region does not code for any gene products and therefore called as the "non-coding" region or "D-loop." As "D-loop" does not contain any gene fragments, it shows nucleotide variability and polymorphisms between individuals. Such differences in the D-loop are possible as it does not code for any gene products that are necessary for the function of the cell. Due to asymmetric distribution of nucleotides, light and heavy strands of mtDNA can be separated in CsCl gradients. The heavy strand (H-strand) contains a high number of Guanine residues in comparison to the L-strand. H strand encodes a total of 28 gene products, whereas, L-strand transcribes eight tRNAs and an enzyme called NADH dehydrogenase 6 (ND6). The structure of mitochondrial genome is illustrated in Fig. 4.4.

The low fidelity of mtDNA polymerase and the apparent lack of mtDNA repair mechanisms are two main causes of high mutation rate in the mitochondrial genome compared with the nuclear genome. Some of the mtDNA regions evolve 5–10 times higher than that of the nuclear genes and show hypervariability due to their high mutation rate. Two hypervariable regions commonly referred as HV1 and HV2 present within the control region of the mitochondrial genome are regarded as suitable for forensic DNA analysis. Besides, a third portion of the control region, called as HV3 is also examined to gather more information on a tested sample. HV3 is found between 438 and 574 bp (Kareem et al. 2016). The HV1 spans between 16,024 and 16,365, whereas, HV2 is located between 73 and 340.

Table 4.4 Comparison between nuclear DNA and mitochondrial DNA in humans

Characteristics	Nuclear DNA	mtDNA
Size	~3.2 billion bp	~16,569 bp
Copies per cell	Fixed (2; 1 allele from each parent)	Variable (hundreds to thousands)
Percentage of total DNA	99.75%	0.25%
Structure	Linear, packed in chromosomes	Circular
Inheritance	Biparental	Uniparental (from mother only)
Chromosomal pairing	Diploid	Haploid
Recombination	Occurs	Does not occur
Replication repair	Occurs	Does not occur
Uniqueness	Unique to one individual (except identical twins and Y lineage)	No uniqueness (same in the maternal lineage)
Mutation rate	Low	5–10 times higher than that of nuclear DNA
Forensic analysis	Analysis of STR markers	Sequencing of hyper variable (HV) regions
Reference sequence	Described by Human Genome Project (2001)	Described by Anderson et al. (1981)

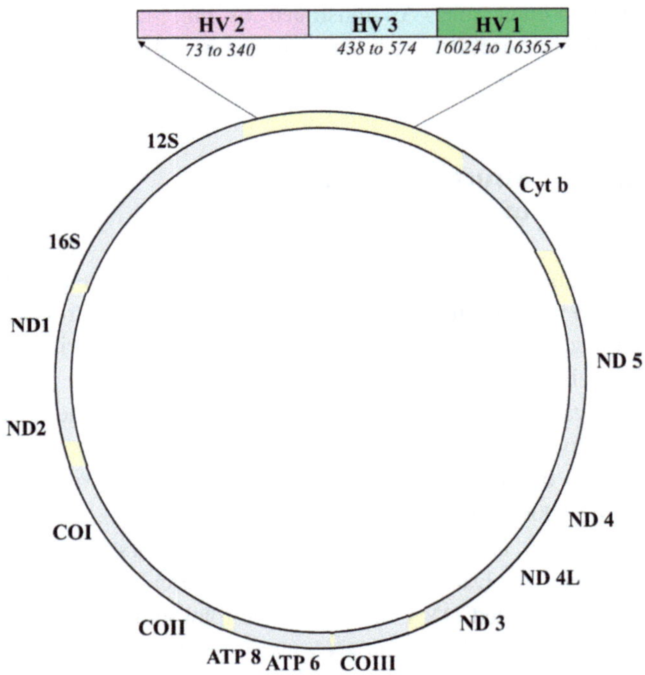

Fig. 4.4 A graphical representation of the double stranded circular mitochondrial genome

Due to the small size of these regions, they are considered to be more suitable for PCR amplification followed by sequencing and have become a great interest for forensic applications. Out of these two hypervariable regions, HV1 is considered as low-resolution region in comparison to HV2, which is a high-resolution region. In these two regions, mutations accumulate more frequently than at any other place in the mitochondrial genome.

4.5.2 Reference Sequence of Human mtDNA

Whole mtDNA was sequenced in Sanger's laboratory in Cambridge, England for the first time in the year 1981 (Anderson et al. 1981). This sequence was obtained from a single individual of European descent. This mtDNA sequence referred as Anderson's sequence or Cambridge Reference Sequence (CRS) is being used to compare with the new sequences. The GenBank accession number of this CRS is M63933. However, the sequencing technique was rudimentary at that time. Thus, the obtained mtDNA sequence contained some HeLa and bovine sequences to fill in gaps. With the advancement of sequencing technologies, the mtDNA was re-sequenced and has been accepted widely as the revised Cambridge Reference Sequence (rCRS). This reanalysis of mtDNA resulted in the 11 discrepancies in the original CRS (Table 4.5). Out of these 11 discrepancies, none of them could be detected in the control region. As only HV1 and HV2 regions are routinely used for forensic DNA analysis, both CRS and rCRS are found to be useful (Andrews et al. 1999). It is available with the NCBI reference sequence NC_012920.1. Another interesting observation of rCRS is that there is a loss of C nucleotide at position 3107 thus, modifying the overall size of the mitochondrial genome to 16,568 bp rather than the originally adopted size of 16,569 bp. A change in the overall numbers of nucleotides would have created the change in position of the other respective nucleotides as well. This would have created a lot of confusion and the previous works on mtDNA sequences would have been difficult to correlate. Thus, the historical numbering has been maintained by including an "N" in place of the 3107 deletion.

4.5.3 Unique Inheritance Pattern of mtDNA

Nuclear DNA shows biparental inheritance. However, mtDNA is maternally inherited. During the formation of zygote, only the sperm's head enters the ovum and merges directly with the egg's nucleus. The fertilized sperm does not contribute any of its parts other than nucleus for the development of the zygote. During the development of blastocyst from the zygote, only the cytoplasm and other cell component of mother's egg cell contribute to mtDNA. The sperm neck and tail contain mitochondria. However, the male mitochondria are generally destroyed either during or shortly after fertilization. Sperm mitochondria disappear during the early embryogenesis stage either by selective destruction, inactivation or dilution (Amorim et al. 2019). Thus, mtDNA sequence of all maternal relatives and siblings is identical barring mutation. As mtDNA does not undergo recombination, maternal relatives of several generations apart can also serve as a reference sample for the identification of an unknown individual. However, recently the biparental inheritance of mtDNA has been reported in 17 members of three multigenerational families and the results were confirmed by two independent laboratories (Luo et al. 2018). A few examples of paternal inheritance of mtDNA are also established in certain animals (Gyllensten et al. 1991).

4.5.4 Heteroplasmy in mtDNA Sequences

Heteroplasmy is the presence of more than one type of mtDNA in an individual. Two or more mtDNA sequences may occur between different cells or within a single cell or within a single mitochondrion. It is now understood that all individuals are heteroplasmic at some level. Some of the heteroplasmy may go undetected due to detection limit of the conventional sequencing technique. The presence of mtDNA heteroplasmy holds special importance in the analysis of forensic evidences such as blood, hair, or skeletal remains. Heteroplasmy is generally regarded as an exceptional event rather than a rule as it is found at minimal frequency in a population. Heteroplasmies are often caused by *de novo* mutations occurring either in the

Table 4.5 A comparative account of Cambridge Reference Sequence (CRS) and Revised Cambridge Reference Sequence (rCRS)

Nucleotide position	Region	CRS	rCRS	Discrepancy
3106–3107	16SrRNA	CC	C	Error
3423	NADH (dehydrogenase 1) [ND1]	G	T	Error
4985	NADH (dehydrogenase 2) [ND2]	G	A	Error
9559	Cytochrome c (oxidase III) [COIII]	G	C	Error
11,335	NADH (dehydrogenase 4) [ND4]	T	C	Error
13,702	NADH (dehydrogenase 5) [ND5]	G	C	Error
14,199	NADH (dehydrogenase 6) [ND6]	G	T	Error
14,272		G	C	Error (inserted bovine sequence)
14,365		G	C	Error (inserted bovine sequence)
14,368		G	C	Error
14,766	Cyt b	T	C	Error (inserted HeLa sequence)

germline or in the somatic tissues. Two types of heteroplasmy are found in mitochondrial DNA, i.e., length heteroplasmy (LHP) and sequence or point heteroplasmy (PHP). LHP mostly occur around the homopolymeric C-stretches in HV1 at positions 16,184–16,193 and HV2 at positions 303–310 (Parson and Bandelt 2007) whereas, sequence heteroplasmy occurs as the presence of two nucleotides at the same position showing overlapping peaks in an electropherogram.

4.5.5 Analysis of mtDNA and Haplotype Group Assignment

In routine forensic mtDNA analysis, sequencing of hyper variable regions is performed. The first step of mtDNA analysis involves the extraction of mtDNA. As there is no specific protocol for extraction of mtDNA exclusively, the routine DNA extraction process is carried out. As mtDNA is present in high copy number per cell, the extraction process needs to be carried out in a clean laboratory environment as mtDNA is more prone to contamination compared to the genomic DNA. Specific quantification of mtDNA is also not available. Thus, routine human DNA quantification is carried out using RT-PCR. Some of the studies have also developed RT-PCR-based quantification of mtDNA (Jackson et al. 2012). But these quantification techniques are still limited to diagnostic purposes. After extraction of mtDNA and quantification, routine sequencing is performed for the HV regions of mtDNA. Sequencing is performed for both forward and reverse primers to generate the contig sequences to obtain a high quality mtDNA sequence. The list of primers used for sequencing of HV1, HV2, and HV3 is given in Table 4.6.

After generating the sequences of D-loop, the obtained sequences are compared with the sequences of rCRS. The variations found in the sequences in comparison to rCRS is noted. For example, if at position 16123 "A" is present in the obtained sequence, whereas, at the same position, "T" is present in the rCRS, the variant can be called as 16123A. If no variation is found in the remaining sequences, it is considered that the remaining sequence is same as that of the rCRS. As similar approach is used for generation of variants in both questioned and reference samples. Then the variants are compared between these two sequences and the results can be interpreted as described in Table 4.7. Furthermore, haplogroups are assigned to each mtDNA sequence. Haplogroup assignment is an additional quality assessment tool for mtDNA dataset. mtDNA haplogrouping can be carried out either manually or through the use of different software such as HaploCart, HaploGrep2, EMMA,

Table 4.6 List of primers used for the sequencing of HV1, HV2, and HV3 of D-loop of mitochondrial DNA

Primers	Sequences (5′ to 3′)	Reference
L15969	CCA AGG ACA AAT CAG AGA AAA AGT C	Hwa et al. (2012)
L15997	CAC CAT TAG CAC CCA AAG CT	
L16262	CTC ACC CAC TAG GAT ACC AAC	
H16401	TGA TTT CAC GGA GGA TGG TG	
L29	GGT CTA TCA CCC TAT TAA CCA C	
H408	CTG TTA AAA GTG CAT ACC GCC A	
H638	GGA CCA AAC CTA TTT GTT TAT GGG	

Table 4.7 General interpretation of mtDNA sequencing results between the questioned and reference samples

Observations	Interpretation
Sequences are fully concordant with common bases at every position	Cannot exclude
Sequences differ at two or more positions	Exclusion
A single unspecified base in one of the sequences; common base at every position	Cannot exclude
Ambiguous bases in both sequences at different positions; common base at every position	Cannot exclude
Heteroplasmic mixture at a position in one sample that is not present in the other; common base at every position (G in both Q and K)	Cannot exclude
Heteroplasmic mixture at the same site in both sequences; common base at every position	Cannot exclude
Sequences identical at every position except one; no indication of heteroplasmy	Inconclusive

MitoTool, HmtDB, mtDNAoffice, and mtDNAmanager. Mitochondrial haplogroup can be defined as a population sharing similar mtDNA sequence. The evolutionary theory of that the origin of modern humans is linked to Africa. Subsequently, they spread around the globe. As per this theory, a hypothetical woman is placed at the root of all these mitochondrial haplogroups commonly called as Mitochondrial Eve. The major haplogroup assigned to this mitochondrial eve is “L” which is most commonly found in Africa. Another macro-haplogroup “M” is found in Asia and America. The descendants of this haplogroups are designated as “C,” “Z,” “D,” “E,” “G,” and “Q.” Another macro-haplogroup “N” is most commonly found in Australia, the Americas, and some parts of Asia. Its descendants can be attributed to haplogroups “N,” “O,” “A,” “S,” “I,” “W,” “X,” and “Y.” Macro-haplogroup R is found mostly in Europe, Northern Africa, the Pacific and parts of Asia and the Americas. Its descendants are haplogroups “B,” “F,” “H,” “V,” “J,” “T,” “U,” and “K” (Loogväli et al. 2004).

4.6 Single Nucleotide Polymorphism (SNPs) in Forensic DNA Analysis

Single nucleotide polymorphism (SNPs) are single nucleotide variations at a unique position. It is the smallest unit of genetic diversity. A one base difference found in at least 5% individuals in a population. A comparative genomic analysis reveals the presence of two or in some cases more than two nucleotides variation at same position in different individuals. Such SNPs are abundant in the human genome and are considered as the most common class of human polymorphism. They are estimated to occur at 1 in every 1000 bases in the human genome. SNPs may be present in the coding as well as non-coding regions of the human genome. SNPs present in the non-coding region are useful as genetic markers in population genetics and evolutionary studies as well as forensic analysis (Gill 2001).

CTTAGCTT 94% in the population (*major allele*)
CTTAGTTT 6% in the population (*minor allele*)

Fig. 4.5 A typical biallelic SNP with designated minor and major allele

4.6.1 Characteristics of SNP Marker

In a biallelic SNP marker, two possible alleles are present in a population. Out of these two possible alleles, one allele is called as minor allele and another one is called a major allele. Those alleles having a frequency of more than 95% are called major alleles, whereas alleles having frequency of <5% are called minor alleles (Fig. 4.5). Though biallelic SNP markers are routinely found in the human genome, triallelic SNPs are rare in nature (Phillips et al. 2020).

4.6.2 SNP Classes

Forensically significant SNPs are categorized into five different classes based on their usefulness and applications. Such different SNP classes include (a) identity informative (ii) SNPs, (b) lineage-informative (li) SNPs, (c) ancestry-informative (ai) SNPs, (d) phenotype-informative (pi) SNPs, and (e) pharmacogenetic SNPs. Forensically relevant SNPs and their selection can be viewed in SNP*for*ID browser (http://spsmart.cesga.es/results.php). These different types of SNPs are described below.

4.6.2.1 Identity Informative SNPs (iiSNPs)

iiSNPs are similar to that of the traditional STRs which are routinely useful for individualization. Such iiSNPs are found in autosomes, X- and Y-chromosomes, as well as within the mitochondrial DNA. For selection of suitable iiSNPs for forensic identification purposes, a similar approach is applied as was taken for the STRs. For selecting iiSNPs, the characteristics such as high heterozygosity, similar allele frequencies in all populations, low F_{ST} value are taken into consideration (Novroski and Cihlar 2022). The examples of some iiSNPs include rs2107612, rs1454361, rs2111980, rs1355366, and rs251934 (Habibi et al. 2019).

4.6.2.2 Lineage-Informative SNPs (liSNPs)

The SNPs that do not undergo recombination and have a low mutation rate can provide useful information regarding the lineage of an individual; such SNPs are called as liSNPs. Such SNPs are highly useful in evolutionary studies as well as kinship analysis. Many such liSNPs have been identified in mitochondrial genome, X- and Y-chromosomes. The uniparental inheritance of liSNP markers found on mitochondrial genome and on the Y-chromosome is the fundamental basis of providing lineage information using liSNPs. However, liSNPs have also been identified in autosomes. Such autosomal liSNPs are neighboring, tightly linked SNPs that are inherited together collectively forming the autosomal haploblocks (Ge et al. 2010). liSNPs present as autosomal haploblocks are currently being used in forensic genetic genealogy practices. A few examples of liSNPs include rs3789669, rs3213892, rs4422272, rs41462, and rs2519154 (Pakstis et al. 2012).

4.6.2.3 Ancestry-Informative SNPs (aiSNPs)

aiSNPs provide information regarding the biogeographic ancestry (BGA) of an individual. Such markers are highly useful in analyzing severely decomposed, skeletonized, mutilated human remains where routine DNA identification is not possible due to unavailability of reference samples for matching. Forensically useful STR markers are highly suitable for identification purpose. However, STR markers are not suitable for prediction of BGA of an individual due to their high mutation rate and high level of locus-specific allele sharing within a population (Novroski and Cihlar 2022). Though liSNPs are also used to prediction of evolutionary history of a population and genetic ancestry of an individual, the uniparental inheritance nature of them does not make them a suitable candidate for accurate prediction of ancestry related information. aiSNPs are selected based on the significance difference in allele frequency across different geographical population. Though many multiplex systems have been developed to derive BGA, some of these include the EUROFORGEN Global AIM-SNP set (Phillips et al. 2014), VISAGE Basic Tool (BT) consisting 112 ancestry predictive SNPs (de la Puente et al. 2021), and Pacifiplex (Santos et al. 2016).

4.6.2.4 Phenotype-Informative SNPs (piSNPs)

piSNPs are the unique type of polymorphic markers that are present in the coding region of the human genome. This is how the piSNPs differ from other types of SNPs which are present either outside the gene or within the intron (non-coding) region of a gene. As per the basic rule of genetics, i.e., genotype codes for phenotype, analysis of genetic variation can be used to predict the phenotype of an individual. Certain external visible characteristics (EVC) such as eye color, hair color, skin color can be predicted using these SNPs thus providing the investigating leads. Despite of the complexity of the human genetic makeup, and phenotype of an individual can be affected by the environmental conditions, many piSNPs have been found useful in predicting the EVC of an individual. Such a branch of forensic DNA analysis is considered as forensic DNA Phenotyping (FDP). Many studies have also used piSNPs to predict the age, height and face shape of an individual. The association of genes with various facial traits of human beings is given in Table 4.8.

4.6.2.5 Pharmacogenetic SNPs

Pharmacogenetic SNPs are the most important and emerging SNP subtypes useful for both forensic genetics and pharmacogenetics. Analysis of such SNPs in postmortem tissue samples can provide useful information regarding the cause and manner of death. In sudden unexplained death cases, analysis of pharamacogenetic SNPs are also found useful. For example, in long QT syndrome (LQTS), mutations in genes encoding cardiac ion channels such as *SCN5A* and *KCNH2* genes leads to ion channel dysfunction. SNPs have been identified in these genes which can be useful in detection the cause of death by the emerging process called as molecular autopsy (Novroski and Cihlar 2022).

Another type of subdivision of SNPs is based on their position in the human genome. Broadly SNPs are divided into two categories, i.e., genic and nongenic. Genic SNPs that occur within a gene are further classified into five different types as described below.

Intronic SNPs: SNPs found within the introns, which are present in the non-coding segment of a gene

5′ UTR SNPs: Such SNPs are found in the 5′ end of a gene which are transcribed into mRNA but not transcribed by the ribosome

3′ UTR SNPs: Such SNPs are found in the 3′ end of a gene which are transcribed into mRNA but not transcribed by the ribosome

Synonymous SNPs: Such a substitution of nucleotide does not change the amino acid

Nonsynonymous SNPs: Such a substitution of nucleotide changes the amino acid

4.6.2.6 Advantages of SNP Markers

The primary advantage of using SNP marker is its small size. A huge quantity of information can be recovered from the degraded samples due to the fact that smaller target region needs to be amplified for an SNP marker. In comparison to that, an array of nucleotides, sometimes hundreds of base pairs are amplified to generate information of an STR marker. Though miniSTRs have been discovered reducing the amplicon size of the STR fragment to 67–123 bp (Coble and Butler 2005). In this regard, SNP markers are thus highly useful in the analysis of degraded samples as extremely small amplicons (50–90 bp) can generate highly useful information (Pontes et al. 2015). Another advantage of the SNP markers are their low mutation rates. The mutation rate of SNPs (10^{-8}) is 100 thousand times lower than that of the STRs (10^{-3}) (Butler et al. 2007). Thus, they are regarded as the more stable markers. The low mutation rate followed by stable nature of inheritance allows an SNP marker to be effectively used in parentage determination, kinship analysis, and identification of mass disaster victims. Many

Table 4.8 Association of genes with forensically relevant phenotypes for FDP applications (Alshehhi et al. 2023)

Sl. No.	Genes	Facial region	Phenotypes	Ancestry
1	TRPC6	Face	Upper and middle facial depth	European
2	ATP8A1	Eye	Upper and lower eyelid sagging severity	European
3	CACNA2D3	Nose	Nose size	African and European
4	LINC00620	Mouth	Mouth morphology measurement and lower lip height	European
5	LINC01965, AHCYP3	Chin/lower face	Chin dimples, jaw slope	Hispanic/Latin American
6	SMG6	Forehead	Forehead protrusion/upper forehead slant angle	Hispanic/Latin American/ East Asian
7	THSD4	Eye	Right eye tail length, outer canthal width	East Asian
8	SFRP2, DCHS2	Nose	Nose roundness, nostril size	Hispanic/Latin American
9	SLC24A5	Mouth	Lip thickness	Hispanic/Latin American
10	CRB1	Chin/lower face	Chin protrusion	Hispanic/Latin American

cases showing ambiguous STR results and cases with high paternity index showing the alleged father may be the biological relative of the real one have been addressed using SNP markers in combination with the STR markers (Phillips et al. 2008; Pinto et al. 2013).

4.6.2.7 Disadvantages of SNP Markers

As most of the SNPs are biallelic in nature, a total of three possible genotypes can be found in the population. For example, if a biallelic SNP has two possible alleles, i.e., C and T, the possible genotypes are CC, CT, and TT. Such limited genetic information obtained from one SNP marker limits its application in forensic DNA based identification. Such limited informativeness is one of the major limiting factors of SNP markers. However, with an increase in the number of SNP markers by multiplexing can increase the overall discrimination power of the SNP multiplex systems. As recommended by Butler et al. (2007), 40–60 numbers of SNPs are required to achieve the same degree of discrimination achieved by 13–15 numbers of STR markers. Though with an increase in the number of SNP markers the overall discrimination power can be increased, multiplexing of ~50 SNPs and analyzing them simultaneously becomes highly difficult without NGS. With the increase in the number of SNP markers, the data interpretation becomes highly challenging and the chance of locus drop out becomes more significant. Besides, another significant disadvantage of using SNP markers is attributed to its less capability in deciphering mixtures.

References

Alshehhi A, Almarzooqi A, Alhammadi K, Werghi N, Tay GK, Alsafar H (2023) Advancement in human face prediction using DNA. Genes (Basel) 14:136. https://doi.org/10.3390/genes14010136

Amorim A, Fernandes T, Taveira N (2019) Mitochondrial DNA in human identification: a review. PeerJ 7:e7314. https://doi.org/10.7717/peerj.7314

Anderson S, Bankier AT, Barrell BG, de Bruijn MHL, Coulson AR, Drouin J, Young IG (1981) Sequence and organization of the human mitochondrial genome. Nature 290:457–465. https://doi.org/10.1038/290457a0

Andrews RM, Kubacka I, Chinnery PF, Lightowlers RN, Turnbull DM, Howell N (1999) Reanalysis and revision of the Cambridge reference sequence for human mitochondrial DNA. Nat Genet 23:147. https://doi.org/10.1038/13779

Ballantyne KN, Keerl V, Wollstein A, Choi Y, Zuniga SB, Ralf A, Vermeulen M, de Knijff P, Kayser M (2012) A new future of forensic Y-chromosome analysis: rapidly mutating Y-STRs for differentiating male relatives and paternal lineages. Forensic Sci Int Genet 6:208–218. https://doi.org/10.1016/j.fsigen.2011.04.017

Brown S, Massilani D, Kozlikin MB, Shunkov MV, Derevianko AP, Stoessel A, Street BJ, Meyer M, Kelso J, Pääbo S, Higham T, Douka K (2022) The earliest Denisovans and their cultural adaptation. Nat Ecol Evolut 6:28–35. https://doi.org/10.1038/s41559-021-01581-2

Budowle B, Adamowicz M, Aranda XG, Barna C, Chakraborty R, Cheswick D, Dafoe B, Eisenberg A, Frappier R, Gross AM, Ladd C, Lee HS, Milne SC, Meyers C, Prinz M, Richard ML, Saldanha G, Tierney AA, Viculis L, Krenke BE (2005) Twelve short tandem repeat loci Y chromosome haplotypes: genetic analysis on populations residing in North America. Forensic Sci Int 150:1–15

Butler JM (2007) Short tandem repeat typing technologies used in human identity testing. Biotechniques 43:ii–v. https://doi.org/10.2144/000112582

Butler JM, Coble MD, Vallone PM (2007) STRs vs. SNPs: thoughts on the future of forensic DNA testing. Forensic Sci Med Pathol 3:200–205. https://doi.org/10.1007/s12024-007-0018-1

Castella V, Gervaix J, Hall D (2013) DIP–STR: highly sensitive markers for the analysis of unbalanced genomic mixtures. Hum Mutat 34(4):644–654. https://doi.org/10.1002/humu.22280

Coble MD, Butler JM (2005) Characterization of new MiniSTR loci to aid analysis of degraded DNA. J Forensic Sci 50:43–53

Dash HR, Rawat N, Das S (2020) Alternatives to amelogenin markers for sex determination in humans and their forensic relevance. Mol Biol Rep 47:2347–2360. https://doi.org/10.1007/s11033-020-05268-y

Dash HR, Rawat N, Vajpayee K, Shrivastava P, Das S (2021a) Useful autosomal STR marker sets for forensic and paternity applications in the central Indian population. Ann Hum Biol 48:37–48. https://doi.org/10.1080/03014460.2021.1877353

Dash HR, Vajpayee K, Srivastava A, Das S (2021b) Prevalence and characterisation of size and sequence-based microvariant alleles at nine autosomal STR markers in the central Indian population. Ann Hum Biol 48:614–620. https://doi.org/10.1080/03014460.2021.2010804

Dash HR, Gautam I, Singh AK, Shrivastava P (2023) Genetic analysis of father-daughter incest using multifaceted STR markers and study of inheritance pattern of alleles. Curr Forensic Sci 1:9–15. https://doi.org/10.2174/2666484401666210809114452

de la Puente M, Ruiz-Ramírez J, Ambroa-Conde A, Xavier C, Pardo-Seco J, Álvarez-Dios J, Freire-Aradas A, Mosquera-Miguel A, Gross TE, Cheung EYY, Branicki W, Nothnagel M, Parson W, Schneider PM, Kayser M, Carracedo A, Lareu MV, Phillips C, On Behalf of the Visage Consortium (2021) Development and evaluation of the ancestry informative marker panel of the VISAGE basic tool. Genes (Basel) 12:1284. https://doi.org/10.3390/genes12081284

Ensenberger MG, Lenz KA, Matthies LK, Hadinoto GM, Schienman JE, Przech AJ, Morganti MW, Renstrom DT, Baker VM, Gawrys KM, Hoogendoorn M, Steffen CR, Martín P, Alonso A, Olson HR, Sprecher CJ, Storts DR (2016) Developmental validation of the PowerPlex® fusion 6C system. Forensic Sci Int Genet 21:134–144. https://doi.org/10.1016/j.fsigen.2015.12.011

Fan H, Chu JY (2007) A brief review of Short tandem repeat mutation. Genomics Proteomics Bioinformatics 5:7–14. https://doi.org/10.1016/S1672-0229(07)60009-6

Fan G, Pan L, Tang P, Zhou Y, Liu M, Luo X (2021) Technical note: developmental validation of a novel 41-plex Y-STR system for the direct amplification of reference samples. Int J Legal Med 135:409–419. https://doi.org/10.1007/s00414-020-02326-9

Ge J, Budowle B, Planz JV, Chakraborty R (2010) Haplotype block: a new type of forensic DNA markers. Int J Legal Med 124:353–361. https://doi.org/10.1007/s00414-009-0400-5

Gill P (2001) An assessment of the utility of single nucleotide polymorphisms (SNPs) for forensic purposes. Int J Legal Med 114:204–210. https://doi.org/10.1007/s004149900117

Gill P et al (2006) New multiplexes for Europe-amendments and clarification of strategic development. Forensic Sci Int 163:155–157

Goedbloed M, Vermeulen M, Fang RN, Lembring M, Wollstein A, Ballantyne K, Lao O, Brauer S, Kruger C, Roewer L, Lessig R, Ploski R, Dobosz T, Henke L, Henke J, Furtado MR, Kayser M (2009) Comprehensive mutation analysis of 17 Ychromosomal short tandem repeat polymorphisms included in the AmpFlSTR Yfiler PCR amplification kit. Int J Legal Med 123:471–482. https://doi.org/10.1007/s00414-009-0342-y

Gopinath S, Zhong C, Nguyen V, Ge J, Lagacé RE, Short ML, Mulero JJ (2016) Developmental validation of the Yfiler(®) Plus PCR Amplification Kit: an enhanced Y-STR multiplex for casework and database applications. Forensic Sci Int Genet 24:164–175. https://doi.org/10.1016/j.fsigen.2016.07.006

Green R, Elliott JL, Norona W, Go F, Nguyen VT, Ge J, Short ML, Mulero JJ, Zhong C (2021) Developmental validation of VeriFiler™ Plus PCR Amplification Kit: a 6-dye multiplex assay designed for casework samples. Forensic Sci Int Genet 53:102494. https://doi.org/10.1016/j.fsigen.2021.102494

Gyllensten U, Wharton D, Josefsson A, Wilson AC (1991) Paternal inheritance of mitochondrial DNA in mice. Nature 352:255–257. https://doi.org/10.1038/352255a0

Habibi S, Ahmadi A, Behmanesh M, Miri A, Tavallaie M (2019) Evaluation of ten SNP markers for human identification and paternity analysis in Persian population. Iranian J Biotechnol 17:e2148. https://doi.org/10.29252/ijb.2148

Hagelberg E, Gray IC, Jeffreys A (1991) Identification of the skeletal remains of a murder victim by DNA analysis. Nature 352:427–429

Hares DR (2015) Selection and implementation of expanded CODIS core loci in the United States. Forensic Sci Int Genet 17:33–34

Hwa HL, Ko TM, Chen YC, Lin CY, Huang YH, Tseng LH, Su YN, Lee JCI (2012) Sequence polymorphisms of mtDNA HV1, HV2 and HV3 regions in eight population groups living in Taiwan. Aust J Forensic Sci 44:243–252. https://doi.org/10.1080/00450618.2011.650208

Jackson CB, Gallati S, Schaller A (2012) qPCR-based mitochondrial DNA quantification: influence of template DNA fragmentation on accuracy. Biochem Biophys Res Commun 423:441–447. https://doi.org/10.1016/j.bbrc.2012.05.121

Jeffreys AJ, Turnerj M, Debenham P (1991) The efficiency of multilocus DNA fingerprint probes for individualization and establishment of family relationships, determined from extensive casework. Am J Hum Genet 48:824–840

Kareem MA, Abdulzahra AI, Hameed IH, Jebor MA (2016) A new polymorphic positions discovered in mitochondrial DNA hypervariable region HVIII from central and north-central of Iraq. Mitochondrial DNA A DNA Mapp Seq Anal 27:3250–3254. https://doi.org/10.3109/19401736.2015.1007369

Kayser M (2017) Forensic use of Y-chromosome DNA: a general overview. Hum Genet 136:621–635. https://doi.org/10.1007/s00439-017-1776-9

Kimpton CP et al (1994) Evaluation of an automated DNA profiling system employing multiplex amplification of four tetrameric STR loci. Int J Legal Med 106:302–311

Kimpton CP et al (1996) Validation of highly discriminating multiplex short tandem repeat amplification systems for individual identification. Electrophoresis 17:1283–1293

King JL, Woerner AE, Muenzler MK, Kapema KB, Bus MM, Smuts A, Ge J, Budowle B (2022) Assessing sequence variation and genetic diversity of currently untapped Y-STR loci. Forensic Sci Int Rep 6:100298. https://doi.org/10.1016/j.fsir.2022.100298

Kraemer M, Prochnow A, Bussmann M, Scherer M, Peist R, Steffen C (2017) Developmental validation of QIAGEN investigator® 24plex QS kit and investigator® 24plex GO! Kit: two 6-dye multiplex assays for the extended CODIS core loci. Forensic Sci Int Genet 29:9–20. https://doi.org/10.1016/j.fsigen.2017.03.012

Liu H, Li X, Mulero J, Carbonaro A, Short M, Ge J (2016) A convenient guideline to determine if two Y-STR profiles are from the same lineage. Electrophoresis 37:1659–1668. https://doi.org/10.1002/elps.201500566

Liu Y, Guo L, Jin H, Li Z, Bai R, Shi M, Ma S (2017) Developmental validation of a 6-dye typing system with 27 loci and application in Han population of China. Sci Rep 7:4706. https://doi.org/10.1038/s41598-017-04548-1

Loogväli EL, Roostalu U, Malyarchuk BA, Derenko MV, Kivisild T, Metspalu E, Tambets K, Reidla M, Tolk HV, Parik J, Pennarun E, Laos S, Lunkina A, Golubenko M, Barać L, Peričić M, Balanovsky OP, Gusar V, Khusnutdinova EK, Stepanov V, Puzyrev V, Rudan P, Balanovska EV, Grechanina E, Richard C, Moisan JP, Chaventré A, Anagnou NP, Pappa KI, Michalodimitrakis EN, Claustres M, Gölge M, Mikerezi I, Usanga E, Villems R (2004) Disuniting uniformity: a pied Cladistic canvas of mtDNA Haplogroup H in Eurasia. Mol Biol Evolut 21:2012–2021. https://doi.org/10.1093/molbev/msh209

Ludeman MJ, Zhong C, Mulero JJ, Lagacé RE, Hennessy LK, Short ML, Wang DY (2018) Developmental validation of GlobalFiler™ PCR amplification kit: a 6-dye multiplex assay designed for amplification of casework samples. Int J Legal Med 132:1555–1573. https://doi.org/10.1007/s00414-018-1817-5

Luo S, Valencia CA, Zhang J, Lee NC, Slone J, Gui B, Huang T (2018) Biparental inheritance of mitochondrial DNA in humans. Proc Natl Acad Sci U S A 115: 13039–13044. https://doi.org/10.1073/PNAS.1810946115

Novroski NMM, Cihlar JC (2022) Evolution of single-nucleotide polymorphism use in forensic genetics. WIRE Forensic Sci 4:e1459. https://doi.org/10.1002/wfs2.1459

Pakstis AJ, Fang R, Furtado MR, Kidd JR, Kidd KK (2012) Mini-haplotypes as lineage informative SNPs and ancestry inference SNPs. Eur J Hum Genet 20: 1148–1154. https://doi.org/10.1038/ejhg.2012.69

Parson W, Bandelt HJ (2007) Extended guidelines for mtDNA typing of population data in forensic science. Forensic Sci Int Genet 1:13–19. https://doi.org/10.1016/j.fsigen.2006.11.003

Pereira L, Prata MJ, Amorim A (2002) Mismatch distribution analysis of Y-STR haplotypes as a tool for the evaluation of identity-by-state proportions and significance of matches–the European picture. Forensic Sci Int 130:147–155

Phillips C, Fondevila M, Magariños MG, Rodriguez A, Salas A, Carracedo A, Lareu MV (2008) Resolving relationship tests that show ambiguous STR results using autosomal SNPs as supplementary markers. Forensic Sci Int Genet 2:198–204. https://doi.org/10.1016/j.fsigen.2008.02.002

Phillips C, Parson W, Lundsberg B, Santos C, Freire-Aradas A, Torres M, Eduardoff M, Børsting C, Johansen P, Fondevila M, Morling N, Schneider P, EUROFORGEN-NoE Consortium, Carracedo A, Lareu MV (2014) Building a forensic ancestry panel from the ground up: the EUROFORGEN Global AIM-SNP set. Forensic Sci Int Genet 11:13–25. https://doi.org/10.1016/j.fsigen.2014.02.012

Phillips C, Amigo J, Tillmar AO, Peck MA, de la Puente M, Ruiz-Ramírez J, Bittner F, Idrizbegović S, Wang Y, Parsons TJ, Lareu MV (2020) A compilation of tri-allelic SNPs from 1000 genomes and use of the most polymorphic loci for a large-scale human identification panel. Forensic Sci Int Genet 46:102232. https://doi.org/10.1016/j.fsigen.2020.102232

Pinto N, Magalhães M, Sousa EC, Gomes C, Pereira R, Alves C, Gusmão L, Amorim A (2013) Assessing paternities with inconclusive STR results: the suitability of bi-allelic markers. Forensic Sci Int Genet 7:16–21. https://doi.org/10.1016/j.fsigen.2012.05.002

Pontes ML, Fondevila M, Laréu MV, Medeiros R (2015) SNP markers as additional information to resolve complex kinship cases. Transfus Med Homother 42: 385–388. https://doi.org/10.1159/000440832

Prieto-Fernandez E, Baeta M, Nunez C, Zarrabeitia MT, Herrera RJ, Builes JJ, de Pancorbo MM (2016) Development of a new highly efficient 17 X-STR multiplex for forensic purposes. Electrophoresis 37:1651–1658. https://doi.org/10.1002/elps.201500546

Purps J, Siegert S, Willuweit S, Nagy M, Alves C, Salazar R, Angustia SM, Santos LH, Anslinger K, Bayer B, Ayub Q, Wei W, Xue Y, Tyler-Smith C, Bafalluy MB, Martinez-Jarreta B, Egyed B, Balitzki B, Tschumi S, Ballard D, Court DS, Barrantes X, Bassler G, Wiest T, Berger B, Niederstatter H, Parson W, Davis C, Budowle B, Burri H, Borer U, Koller C, Carvalho EF, Domingues PM, Chamoun WT, Coble MD, Hill CR, Corach D, Caputo M, D'Amato ME, Davison S, Decorte R, Larmuseau MH, Ottoni C, Rickards O, Lu D, Jiang C, Dobosz T, Jonkisz A, Frank WE, Furac I, Gehrig C, Castella V, Grskovic B, Haas C, Wobst J, Hadzic G, Drobnic K, Honda K, Hou Y, Zhou D, Li Y, Hu S, Chen S, Immel UD, Lessig R, Jakovski Z, Ilievska T, Klann AE, Garcia CC, de Knijff P, Kraaijenbrink T, Kondili A, Miniati P, Vouropoulou M, Kovacevic L, Marjanovic D, Lindner I, Mansour I, Al-Azem M, Andari AE, Marino M, Furfuro S, Locarno L, Martin P, Luque GM, Alonso A, Miranda LS, Moreira H, Mizuno N, Iwashima Y, Neto RS, Nogueira TL, Silva R, Nastainczyk-Wulf M, Edelmann J, Kohl M, Nie S, Wang X, Cheng B et al (2014) A global analysis of Y-chromosomal haplotype diversity for 23 STR loci. Forensic Sci Int Genet 12:12–23

Purps J, Geppert M, Nagy M, Roewer L (2015) Validation of a combined autosomal/Y-chromosomal STR approach for analyzing typical biological stains in sexual-assault cases. Forensic Sci Int Genet 19:238–242

Roewer L (2019) Y-chromosome short tandem repeats in forensics-sexing, profiling, and matching male DNA. WIREs Forensic Sci 1:e1336. https://doi.org/10.1002/wfs2.1336

Santos C, Phillips C, Fondevila M, Daniel R, van Oorschot RAH, Burchard EG, Schanfield MS, Souto L, Uacyisrael J, Via M, Carracedo A, Lareu MV (2016) Pacifiplex: an ancestry-informative SNP panel centred on Australia and the Pacific region. Forensic Sci Int Genet 20:71–80. https://doi.org/10.1016/j.fsigen.2015.10.003

Satoh M, Kuroiwa T (1991) Organization of multiple nucleoids and DNA molecules in mitochondria of a human cell. Exp Cell Res 196:137–140

Scherer M, König M, Bussmann M, Prochnow A, Peist R (2015) Development and validation of the new investigator® Argus X-12 QS kit. Forensic Sci Int Genet Suppl Ser 5:e256–e257. https://doi.org/10.1016/j.fsigss.2015.09.102

Stanley UN, Khadija AM, Bukola AT, Precious IO, Davidson EA (2020) Forensic DNA profiling: autosomal Short tandem repeat as a prominent marker in crime investigation. Malaysian J Med Sci 27(4): 22–35. https://doi.org/10.21315/mjms2020.27.4.3

Star B, Hansen MHS, Skage M, Bradbury IR, Godiksen JA, Kjesbu OS, Jentoft S (2016) Preferential amplification of repetitive DNA during whole genome sequencing library creation from historic samples. Sci Technol Archaeol Res 2:36–45. https://doi.org/10.1080/20548923.2016.1160594

Sullivan KM, Mannucci A, Kimpton CP, Gill P (1993) A rapid and quantitative DNA sex test: fluorescence-based PCR analysis of X-Y homologous gene amelogenin. BioTechniques 15:636–638

Szibor R et al (1997) Mitochondrial D-loop 3′ (CA)n repeat polymorphism: optimization of analysis and population data. Electrophoresis 18:2857–2860

Teama S (2018) DNA polymorphisms: DNA-based molecular markers and their application in medicine. In: Liu Y (ed) Genetic diversity and disease susceptibility. IntechOpen, London. https://doi.org/10.5772/intechopen.72269

Thangaraj K, Reddy AG, Singh L (2002) Is the amelogenin gene reliable for gender identification in forensic casework and prenatal diagnosis? Int J Legal Med 116:121–123

Thompson JM, Ewing MM, Frank WE, Pogemiller JJ, Nolde CA, Koehler DJ, Shaffer AM, Rabbach DR, Fulmer PM, Sprecher CJ, Storts DR (2013) Developmental validation of the PowerPlex® Y23 system: a single multiplex Y-STR analysis system for casework and database samples. Forensic Sci Int Genet 7:240–250. https://doi.org/10.1016/j.fsigen.2012.10.013

Vraneš M, Scherer M, Cornelius S, Willuweit M, Kraemer M, Terramagra D, Kohns A, Prochnow A (2022) Developmental validation of the Investigator Argus Y-28 QS kit. Forensic Sci Int Genet Suppl Ser 8:274–275. https://doi.org/10.1016/j.fsigss.2022.10.060

Wei T, Liao F, Wang Y, Pan C, Xiao C, Huang D (2018) A novel multiplex assay of SNP-STR markers for forensic purpose. PLoS One 13(7):e0200700. https://doi.org/10.1371/journal.pone.0200700

Yuan Y, Ju YS, Kim Y, Li J, Wang Y, Yoon CJ, Yang Y, Martincorena I, Creighton CJ, Weinstein JN, Xu Y, Han L, Kim HL, Nakagawa H, Park K, Campbell PJ, Liang H, PCAWG Consortium (2020) Comprehensive molecular characterization of mitochondrial genomes in human cancers. Nat Genet 52:342–352. https://doi.org/10.1038/s41588-019-0557-x

Zhang YD, Shen CM, Meng HT, Guo YX, Dong Q, Yang G, Yan JW, Liu YS, Mei T, Huang RZ, Zhu BF (2016) Allele and haplotype diversity of new multiplex of 19 ChrX-STR loci in Han population from Guanzhong region (China). Electrophoresis 37:1669–1675. https://doi.org/10.1002/elps.201500425

5 Fast, High-Sensitive, and High-Resolution DNA Techniques

5.1 Introduction

Upon the creation of the International Olympic Committee, the motto for the Olympic was proposed by Pierre de Coubertin. The adopted Olympic motto in the latinized version is "hendiatris *Citius*, *Altius*, *Fortius*" which can be translated as "*Faster*, *Higher*, *Stronger*." Criminal justice system expects an array of techniques with similar characteristics for the analysis of forensic exhibits to aid in the process of justice delivery. Most of the cases pertaining to the examination of biological samples are attributed to forensic DNA examination. The present-day DNA analysis techniques have seen much progress in terms of its reliability, robustness, sensitivity, and faster processing time. When the technique was invented in the year 1985 by Sir Alec Jeffreys, single or multilocus probes were in practice. Similarly, when the first short tandem repeat (STR) kit was launched by Promega Corporation in 1994, only three STR loci were included in the multiplex system, i.e., CSF1PO, TPOX, and TH01 which was popularly called as CTT triplex. Gradually, the number of STR markers included in the multiplex systems increased and now-a-days, most of the STR multiplex kits have more than 20 STR markers. Inclusion of such a high number of STR markers provide high resolution to the technique by generating a high discrimination power, low matching probability, and many other useful forensic and paternity parameters.

During the inception of forensic DNA analysis, the RFLP technique coupled with variable number of tandem repeat (VNTR) analysis was in the requirement of a huge quantity of DNA and the processing time was very high. However, with the advent of advanced molecular techniques and the inception of STR and SNP markers followed by the adoption of capillary electrophoresis (CE), and next generation sequencing (NGS) techniques, the processing time as well as DNA quantity requirement has been further decreased. Over the last few decades, the use of automation techniques has gained popularity in forensic DNA analysis and have decreased the human intervention drastically. This has not only decreased the chance of human error and contamination level, but also contributes significantly in reducing the processing time of the routine forensic DNA analysis workflow. Recently the portable all-in-one systems have also been discovered and validated for their use in forensic case works. Besides reducing the processing time, such instruments have been found highly useful in case work analysis where the samples are tested in the field. Such tests can be considered homologous to the point of care tests that have gained popularity in the clinical setup. In this regard, the current chapter discusses the significant advancements in the field of forensic DNA analysis in terms of the sensitivity, reliability, robustness, speed, and resolution of the technique.

H. R. Dash et al., *Advancements in Forensic DNA Analysis*,
https://doi.org/10.1007/978-981-99-6195-5_5

5.2 Sensitivity of the Advanced Techniques

Forensic Science is regarded as a science of comparison. Questioned samples are generally compared with the reference samples and a conclusion is made. For forensic DNA analysis, the routine reference samples are collected mostly in the form of blood or buccal swabs. In either of the forms the obtained DNA quantity is sufficiently high. However, the questioned samples are highly challenging to process many a times. In many instances, the quantity of biological materials found in the crime scene is much less. Sometimes they are not visible to the naked eye also. In this regard, the chance of getting a sufficient quantity of DNA from the biological samples becomes challenging. The characteristic difference between questioned sample and reference sample is given in Table 5.1.

During the inception of forensic DNA analysis, VNTR-based Restriction Fragment Length Polymorphism (RFLP) technique was employed in 1985. This primitive technique of forensic DNA analysis requires not less than 50 ng of input DNA quantity (Budowle et al. 1994). The invention of Polymerase Chain Reaction (PCR) technique by Karry Mullis has revolutionized the forensic DNA analysis and is considered to be the most significant discovery in forensic science in last 50 years (Kaunitz 2015). Immediately after the discovery of PCR technique, the first PCR-based DNA test kit started to appear on the market. One of such kit was for targeted amplification of Human Leukocyte Antigen (HLA) DQA1 gene present on the chromosome 6 (Comey and Budowle 1991). In mixture analysis, AmpliType HLA-DQ Alpha test kit showed a detection sensitivity of at least 0.2%, which is much better than the 10% detection sensitivity in non-PCR techniques. The first VNTR marker that was amplified successfully was D1S80 in the year 1993. This is a minisatellite present on chromosome no. 1 having repeat unit of 16-bp and PCR product ranging from 400–800 bp. Due to the ease of amplification, this technique required as less as 0.1 ng of genomic DNA (Kloosterman et al. 1993). Hence, the invention of PCR technique coupled with forensic DNA analysis significantly reduced the DNA quantity requirement.

Gradually, the technique advanced further and the STR markers are widely accepted as the form of genetic polymorphism for their use in forensic DNA analysis. The first multiplex system consisting of HUMCSF1PO, HUMTPOX, and HUMTH01 loci popularly called as CTT triplex was having a sensitivity rate of 0.25 ng of template DNA (Micka et al. 1996). This multiplex PCR dependent system showed a significant decrease in the amount of DNA required for forensic DNA analysis. Later on, the number of STR markers included in the various kits increased further. In 1999, it showed the development of Second-Generation Multiplex (SGM) plus kit consisting of ten STR markers by Applied Biosystems. Though the sensitivity of this kit was further increased; only 94% of the profiles could be generated using ≥200 pg/μL of the template DNA (Castella et al. 2004). With the implementation of Combined DNA Index System (CODIS) core loci in 1997 by Federal Bureau of Investigation (FBI), new multiplex systems started to become available commercially. One such kit having 13 CODIS markers is the PowerPlex® 16 HS system. The developmental validation of this kit found the optimum DNA input required for the appropriate analytical interpretation of this system is 0.25–2.0 ng (White et al. 2015). Another kit, i.e., AmpFlSTR® Identifiler® Plus showed the generation of full profile with a 29-cycle PCR protocol using 0.062 ng of DNA. However, the use of 125 pg of DNA showed the generation of a complete profile with both 28 and 29 cycle of PCR condition (Wang et al. 2012).

Gradually, with the inception of expanded-CODIS since 1st January 2017, new STR multiplexing kits have been introduced which are most commonly used in various forensic science laboratories. In this regard Elwick et al. (2018) compared the sensitivity of two new generation kits, i.e., GlobalFiler® kit and Investigator® kit, both having at least 20 expanded-CODIS loci. This study suggested that the sensitivity of GlobalFiler® kit is slightly better than that of the Investigator® kit. Complete STR profiles were

Table 5.1 Characteristic difference between questioned sample and reference sample

Characteristics	Questioned sample	Reference sample
Sample source	Unknown	Known
DNA quantity	May be high or low	Mostly high
DNA quality	May be good or bad	Mostly good
Sample source	May be single source or mixed source	Always single source
Sample collection	Collected from the environment or body as it is	Collected and stored in a controlled manner
Contamination	Exogenous contamination might be present	Exogenous contamination is absent
Example	Blood-stained soil, clothing, vaginal swab, bone, teeth	Peripheral blood, buccal swab

obtained for all the samples using both the kits at an input DNA concentration of 250 pg. However, using 125 pg DNA template, 100% alleles were recovered in GlobalFiler® kit and 91% alleles were recovered in Investigator® kit (Elwick et al. 2018). Similarly, another post-Expanded Locus kit PowerPlex® Fusion 6C System also exhibits a high sensitivity of 125 pg (Ensenberger et al. 2016).

Besides autosomal STR kits, many commercially available kits are also found to amplify Y- and X-STRs. The Y-STR multiplex kits have shown tremendous sensitive to the input DNA concentration and most of the kits generate a complete Y-STR profile with as low as 62.5 pg of input DNA. The Yfiler® Plus Amplification Kit which amplifies 27 Y chromosomal STR markers generates a complete profile with 70 pg of input DNA (Andersen et al. 2017). Another 5-dye multiplex system capable of simultaneous amplification of 23 Y-STRs, i.e., PowerPlex® Y23 system showed a greater sensitivity with the amplification capability from as little as 62.5 pg of input DNA concentration (Gehrig and Castella 2013). A comparative analysis of three Y-STR multiplex kits also showed that Yfiler® Plus PCR Amplification Kit (Applied Biosystems) and PowerPlex® Y23 (Promega®) System showed a better sensitivity than AmpFLSTR® Yfiler™ PCR Amplification Kit (Applied Biosystems) by amplifying Y-STRs using as low as 0.5 ng of input DNA concentration in sexual assault cases (Ferreira-Silva et al. 2018). A relatively new Y-STR multiplex kit, i.e., Argus Y-28 QS kit has the potential to amplify 27 Y-STRs simultaneously with as little as 100 pg input DNA both in presence and absence of female DNA background (Vraneš et al. 2022). A multiplex X-STR kit MentypeRArgusX-UL showed the sensitivity of 100 pg of input DNA (Castella et al. 2006). The most commonly used Investigator Argus X-12 QS Kit has a sensitivity of as low as 125 pg, whereas, the recommended DNA input is 500 pg for the optimum result (Scherer et al. 2015).

5.3 Cell Selection and Cell Picking Methods

In routine biological samples, such as sexual assault evidences experience a huge challenge to the forensic DNA analysts. Such samples are mostly of mixed source. In sexual assault evidence, many factors govern the final outcome of a DNA report. Though the sampling time is recommended for most of the sexual assault related evidence types (Table 5.2), in many instances the evidence samples are collected inappropriately. As the date of sampling and the date of incidence differs largely, the M:F ratio is increased further (Tozzo et al. 2018). In such cases use of highly sensitive kits also cannot generate a suspect's DNA profile. For selective extraction of DNA from sperm cells in a mixture of epithelial and sperm cells, preferential lysis method developed by Gill et al. (1985) has been widely used in the forensic DNA analysis laboratories. This technique is time consuming and requires multiple sample manipulation steps which may lead to the loss of critical sperm. In order to overcome such problem, various targeted

Table 5.2 Recommended sampling time in sexual assault evidences for optimum DNA outcome

Samples	Ideal time for sampling		References
Oral	Within 6 h of sexual contact		Anderst (2011), Magalhães et al. (2015)
Rectum	Within 24 h of sexual contact		
Vagina	Within 72 h of sexual contact		
Samples	**Maximum reported recovery time**		**References**
	Motile sperm	**Non-motile sperm**	
Vagina	6–28 h	14 h to 10 days	Siegel et al. (2013)
Cervix	3–7 days	7.5 to 19 days	
Mouth	–	2 h to 1.5 days	
Rectum	–	4 h to 5 days	
Anus	–	2 h to 2 days	
Dried vaginal secretion stain	–	1 year	Magalhães et al. (2015)

cell selection and cell picking techniques have been developed. Some of the techniques are been discussed below.

5.3.1 Laser Capture Microdissection Technique

Laser Capture Microdissection (LCM) is an established technique for selective isolation of cells from a biological mixture when they are present in very low concentration. This technique allows direct visualization, dissection and recovery of specific cells and tissue from microscope slides using laser illumination. Laser energy is used to increase the specific signal of a targeted cell in comparison to the signal from the whole tissue. LCM involves putting a layer of transparent ethylene vinyl acetate to the slide containing a mixture of cells. Furthermore, a low-power laser beam is directed at the film which extends the long chain polymers of the film, which facilitates the capturing of cells or substructure of interest. Mostly, laser energy of infrared (810 nm) or ultraviolet (355 nm) or both in combination are used (Golubeva et al. 2013). The advantage of this technique is that, after LCM, the isolated cells can be subjected to DNA, RNA, and protein analysis.

DNA from a single spermatozoon on slide which has already been stained with hematoxylin and eosin was captured using LCM method. The DNA from that single spermatozoon was amplified using the improved primer extension preamplification (I-PEP) polymerase chain reaction (PCR) method to generate a complete STR profile successfully (Miyazaki et al. 2008). Anslinger et al. (2007) marked male and female cells categorically in a mixture of cells using a fluorescent X/Y-probe CEP X Spectrum Orange/Y Spectrum Green DNA probe kit. Furthermore, these cells were extracted using LCM and DNA profiling was carried out in 20 such LMD isolated, hybridized cells (Anslinger et al. 2007). Low volume PCR technique coupled with LCM have been proved to be better suitable for DNA profiling from spermatozoa (Li et al. 2011). Fluorescence in situ hybridization coupled with laser microdissection has also been proved to be useful in identifying and isolating male cells in azoospermic sexual assault cases (McAlister 2011). An interesting study found that LCM technique outperformed preferential lysis technique in 15 out of 16 mixed source samples. Thus, LCM technique is considered to be the best alternative to available chemical techniques for successful generation of sperm DNA profile in a mixed source samples involving sexual assault cases (Elliott et al. 2003).

5.3.2 Other Techniques

Many techniques have been discovered recently for the efficient separation of sperm cells and the epithelial cells in sexual assault cases. Those alternative techniques separate these two types of cells based on either their different sediment velocities and different adherence properties (Horsman et al. 2005), size-based acoustic trapping of sperm cells in epithelial cell (EP) lysate (Norris et al. 2009), filtration or dielectrophoresis (DEP) (Chen et al. 1998). However, most of these techniques generate a low pure sample as well as low amount of product. In addition, the advanced techniques such as LCM and antibody-based fluorescence-activated cell sorting (FACS) have been proposed. Though these techniques are highly suitable for the low sperm content samples, these techniques are less cost effective. Recently, microfluidic techniques have been developed for cell separation. Such microfluidic devices rely on either DEP fields, magnetic fields, and sound pressure fields. These microfluidic devices are cheap, disposable to prevent contamination, and they can be integrated with any automated platforms (Liu et al. 2015).

5.4 Tolerance to PCR Inhibitors

PCR has become an integral part of the routine forensic DNA analysis process. Forensic DNA analysis utilizes the PCR technique to amplify the STR markers, mini-STR markers, and hypervariable (HV) regions of mitochondrial DNA for subsequent analysis. Despite having a good quantity of DNA, PCR inhibition is the major cause of PCR failure in forensic DNA analysis. Though PCR process is widely used in different molecular biology experiments, problems of PCR inhibitors are not that much experienced in comparison to the forensic biological samples. Forensic biological samples are mostly recovered from the environment. Such samples are exposed to harsh environmental conditions. Some of these biological samples contain some inherent molecules/compounds which have the potential to act as PCR inhibitors. Some of the important inherent molecules are hemoglobin in blood, calcium in bone, humic acid in soil (Sidstedt et al. 2018). Such molecules are coextracted in DNA and can act as potential PCR inhibitors to spoil the PCR process. Besides as forensic biological samples are exposed to harsh environmental conditions, they may be contaminated with many other biological entities like microorganisms. The by-products of such contaminated biological entities present in the forensic samples can also act as potential PCR inhibitors. PCR inhibitors are a heterogenous group of chemicals. A single molecule or a mixed source of molecules may be found in different samples. Similarly, the nature of the compounds also varies greatly, and they may be organic compounds or inorganic compounds. These group of compounds act on different components of PCR process, i.e., the template DNA, the nucleotides, the amplification primers, Mg^{2+} and the polymerase enzyme (Alaeddini 2012).

To facilitate DNA profiling from the biological samples having higher amount of PCR inhibitors, the DNA extraction technique should be chosen selectively. The DNA extraction technique having the properties of maximized DNA recovery, removal of inhibitors, removal or inhibition of nucleases, and maximization of the quality of DNA can be considered as a gold standard technique to be used in forensic DNA analysis (Miller et al. 1999). A study compared the four commonly used DNA extraction technologies, i.e., PowerClean® DNA Clean-Up kit, DNA IQ™ System, Phenol-Chloroform extraction, and Chelex®-100 methods to study their efficacy in the removal of potential PCR inhibitors from the samples. Eight commonly encountered PCR inhibitors were included in this study, i.e., melanin, humic acid, collagen, bile salt, hematin, calcium ions, indigo, and urea. It showed that each of the PCR inhibitors was completely removed by PowerClean® DNA Clean-Up kit and DNA IQ™ System. However, Phenol-Chloroform method and Chelex®-100 methods are not suitable to remove the potential PCR inhibitors from the samples (Hu et al. 2015). Various DNA extraction protocols are used in forensic DNA analysis laboratories. They may be automated or manual techniques. Based on the chemistry used for DNA extraction, they may be organic methods, column-based methods or magnetic beads-based methods. Out of all the techniques used, magnetic beads-based techniques are considered to be the gold standard techniques in terms of removal of PCR inhibitors (Bordelon et al. 2013).

Most of the new generation autosomal STR kits are robust and show significant inhibitor tolerance. As per the guidelines of International Society of Forensic Genetics (ISFG), the multiplex STR kits should be validated for their inhibitor tolerance toward the commonly encountered potential PCR inhibitors. The commonly encountered PCR inhibitors tested during developmental validation of a multiplex STR kit include heme, humic acid, EDTA, melanin, Ca^{2+}, and tannin (Huang et al. 2022). Out of the three most commonly used autosomal STR kits, 24plex QS Kit (QIAGEN), the GlobalFiler™ PCR Amplification Kit (Applied Biosystems™), and the PowerPlex® Fusion 6C System (Promega), Investigator 24plex showed a higher tolerance to common PCR inhibitors such as humic acid and tannic acid (Lin et al. 2017).

5.5 DNA Profiling in Degraded Samples

Besides PCR inhibitors, routine forensic DNA analysis experiences a huge problem from the degraded samples. As most of the forensic biological samples are exposed to harsh environmental samples, they experience degradation. Degradation of the DNA may be experienced in terms of fragmentation of DNA. DNases degrade the chromosomal DNA into two steps, i.e., first into large, i.e., 50–100 kb followed by into nucleosomal units (Kawane et al. 2014). Various biological factors such as bacteria, fungi, and enzymes as well as environmental factors such as sunlight, heat, and humidity, can facilitate the rate of DNA degradation (Garvin et al. 2013). Weather conditions in an outdoor crime scene such as season, rainfall also affect greatly on the quality of DNA. In routine forensic DNA analysis, small sized STR markers are used. Most of the commercially available kits used now-a-days target these STR markers of ~100 to ~400 bp size. Hence, to generate a complete STR profile, template DNA should be as intact as possible.

However, use of mini-STRs has been recommended for the generation of a complete profile in degraded samples. Mini-STR primers can magnify the STR loci to increase the chance of STR amplification from the degraded samples. Thus, using mini-STR technology a complete DNA profile can be generated in compromised samples. The quality of DNA profiling result is also increased many folds by using the mini-STR technology. Using mini-STR technology a complete profile can be obtained from STR loci with allele size <150 bp. Mini-STR markers relies on the increase of proximity of the forward and reverse primers to the STR region (Fig. 5.1). In such scenario, the amplicon size in significantly reduced and large genetic information can be recovered from these degraded biological samples. Commercially available AmpFl STR™ MiniFiler™ PCR amplification kit amplifies eight autosomal loci, i.e., D13S317, D7S820, D2S1338, D21S11, D16S539, D18S51, CSF1PO, FGA, and the sex determining marker amelogenin (Mulero et al. 2008). The amplicon size is reduced significantly in this kit, i.e., D7S820 (129 bp), D13S317 (99 bp), D21S11 (33 bp), D2S1338 (183 bp), D18S51 (168 bp), D16S539 (157 bp), FGA (87 bp), and CSF1PO (201 bp). A drawback is that more dyes are needed to analyze set of STR markers.

5.6 Number of STR Markers in the New Generation Multiplex Systems

After the completion of Human Genome Project in 1997, it was found that human genome harbors over 20,000 tetra nucleotide STR markers. Tetra nucleotide STR markers are preferred for forensic DNA analysis because of their small size and high heterozygosity and stand out performance in other forensic and paternity parameters. The use of product rule of statistics facilitates the higher combined discrimination power of the multiple STR markers when analyzed simultaneously. In this regard, a core set of 13 STR loci were selected by the Federal Bureau of Investigation (FBI) Laboratory in November 1997, collectively

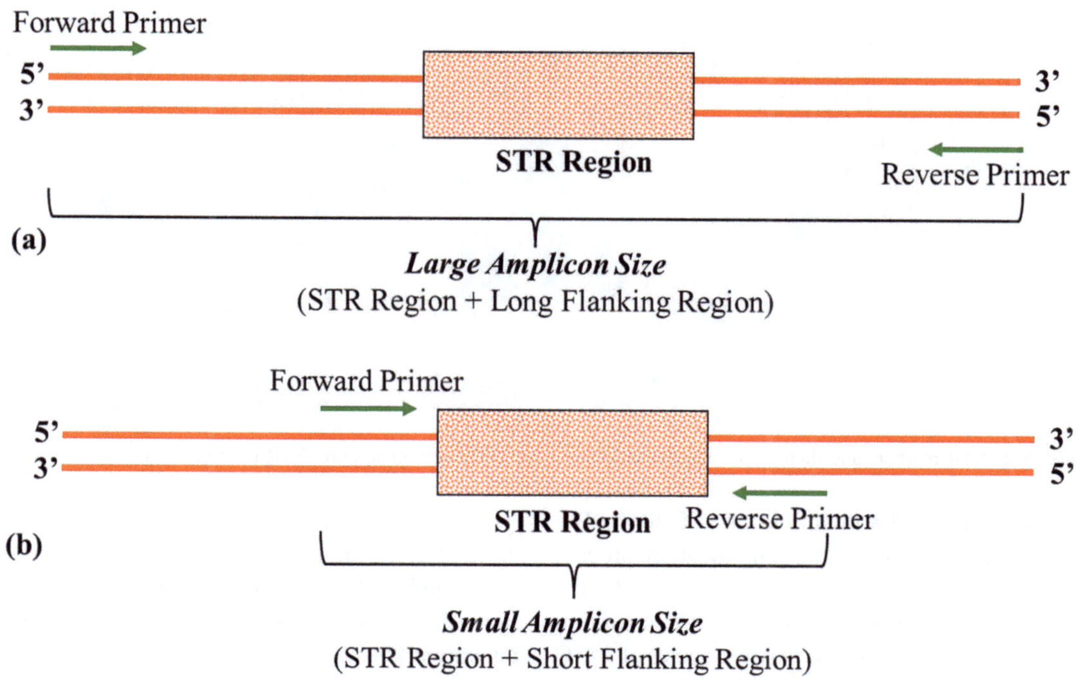

Fig. 5.1 Selection of primers in (**a**) conventional STR analysis, (**b**) in small or mini STR analysis

called as core CODIS loci. However, since 1st January, 2017, the number of STR loci were further increased to 20 to further increase the discrimination power of a generated STR profile (Hares 2015). These set of loci are now called as expanded-CODIS loci. The basis of expansion of the core STR loci from 13 to 20 was (1) to reduce the likelihood of adventitious matches, (2) to increase international compatibility and assisting law enforcement data-sharing efforts, and (3) to increase the discrimination power.

In line with the increase in core STR loci, the commercially available kits have also increased the number of STR markers to more than 20 in the respective multiplex systems. Most of these new generation STR multiplex kits are 6-dye enabled kits as discussed in chapter 1. Different markers have been grouped into five sets and labeled with labeled with one dye each. The sixth dye is reserved for the size standard. The list of markers and their respective labeled dye is shown in Fig. 5.2. A recent study showed that, with the increase in the number of STR markers in a multiplex system, decreases the cumulative match probability of a DNA profile. This study demonstrated various forensic and paternity parameters of various sets of STR markers in the central Indian population. A significant decrease in the matching probability was found from 2.2×10^{-15} to 1.11×10^{-23} when the number of STR markers was increase from 13 CODIS to 20 expanded-CODIS (Dash et al. 2021a). This shows that with the increase in the number of STR markers, the cumulative forensic and paternity parameters of a STR profile gains superiority.

5.7 Fast Amplification Conditions

As the forensic DNA analysis is moving toward generating a faster result, the PCR amplification time is reduced in the new generation multiplex kits. The older kits like AmpFℓSTR™ Identifiler™ PCR Amplification Kit (Thermo Scientific) and PowerPlex® 16 HS System (Promega) showed higher amplification times for less STR markers. The PCR conditions of Identifiler kit is an initial denaturation step of 95 °C for 11 min followed by 28 cycles of 94 °C for 1 min, 59 °C for 1 min, 72 °C for 1 min, followed by a final extension of 60 °

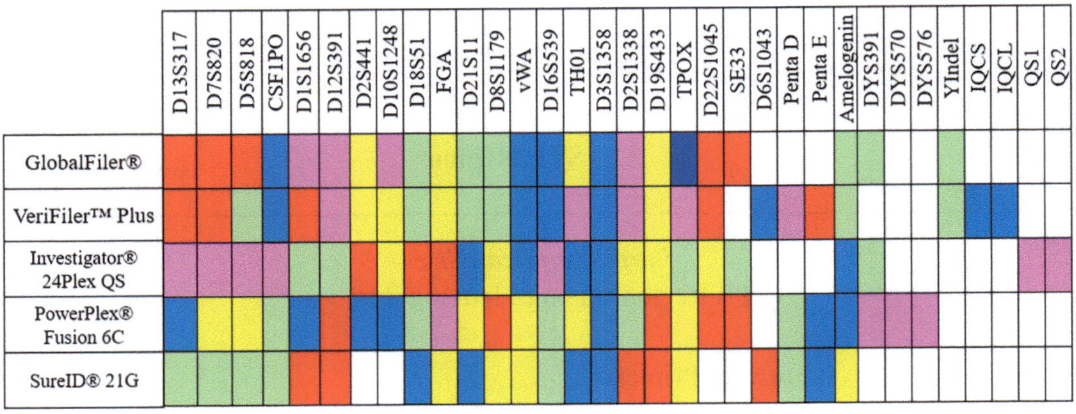

Fig. 5.2 STR markers and their respective dye colors present in the new generation STR multiplex kits

C for 60 min and a final hold of 4 °C for ⍰. The overall completion of the amplification process takes around 155 min to complete. Similarly, the PowerPlex® 16 HS System has a PCR condition of 96 °C for 2 min followed by ten cycles of 94 °C for 30 s, 60 °C for 30 s, 70 °C for 45 s, followed by 22 cycles of 90 °C for 30 s, 60 °C for 30 s, 70 °C for 45 s, followed by a final extension of 60 °C for 30 min and a final hold at 4 °C for ⍰. The whole process takes around 88 min to complete. Such a PCR process time is very high considering the number of STR markers to be amplified.

In contrast to that, the post-expanded-CODIS STR multiplex kits have significantly reduced amplification time. The PowerPlex® Fusion 6C coamplifies 27 STR loci. The PCR conditions for simultaneous amplification of 27 loci is 96 °C for 1 min, then 29 cycles of 96 °C for 5 s and 60° for 1 min followed by 60 °C for 10 min and a final hold of 4° for ⍰. The whole process takes around 43 min to complete. Similarly, another new generation kit Globalfiler has the PCR condition as 95 °C for 1 min, followed by 30 cycles of 94° for 10 s, 59 °C for 90 s, followed by final extension of 60° for 10 min and a final hold of 4° for ⍰. The overall process takes around 61 min to complete. In addition to that, another new generation multiplex kit Investigator® 24plex QS (Qiagen) has the PCR protocol as 3 cycles of 98 °C for 30 s, 64 °C for 55 s, and 72 °C for 5 s, followed by 27 cycles of 96° for 10 s, 61 °C for 55 s, 72 °C for 5 s, followed by 68 °C for 5 min and 60 °C for 5 min and a final hold of 4° for ⍰. This amplification process takes around 46 min to complete. This shows that the significant reduction in PCR process to under 1 h to complete in comparison to the more than 1 h protocol for most of the old generation kits.

Besides a significant reduction in PCR time, most of the new generation kits have developed the additional features of direct amplification from the samples. Such samples, mostly the reference samples taken on blood collection cards can be processed directly for PCR using those kit without extraction and quantification of DNA. This ultimately reduces the processing time of the samples and a DNA profile can be generated directly from the samples. A few examples of such kits include AmpFlSTR® Identifiler® Direct kit (Thermo Scientific), GlobalFiler™ Express kit (Thermo Scientific), Investigator 24plex GO! Kit (Qiagen) and many others (Dash et al. 2020). Besides, the microchip-based systems are gaining popularity for the rapid and small volume amplification of STR markers which shows a promising futuristic application of these techniques (Cornelis et al. 2018).

5.8 Faster and Reliable DNA Quantification Techniques

Owing to the exposure to the adverse environmental conditions, the DNA content of the biological samples varies a lot. Besides the quantity of DNA, the compromised forensic samples are highly prone to contamination and degradation. Thus, prior to the downstream processing, the extracted DNA samples should be checked for the quantity as well as quality of the extracted DNA. Many techniques are available for the quantification of DNA in the forensic samples. The basic techniques of DNA quantification include UV spectrometry, gel-based techniques, dye staining, and blotting techniques. Several modifications have been carried out using these basic techniques to generate a faster reliable result.

One such modification of the basic technique includes the use of NanoDrop Spectrophotometer. It is an extension and improvement of the UV Spectrophotometer technique. It explores the basic quantification technique of DNA which quantifies DNA on the basis of absorbance at 260 nm. Besides, the purity of DNA samples can be assessed by using 260/280, 260/230, and 260/325 nm absorbance ratios. The advantage of this technique is the requirement of small volume of samples for measurement. As little as 1 μL of the sample can be measured using this instrument. Despite many advantages, there exists some limitations of this technique. This technique quantifies DNA on the basis of the absorbance of the samples at 260 nm. This wavelength is absorbed by ssDNA, dsDNA, and RNA. Therefore, this method cannot distinguish among some of these types of nucleic acids (Bruijns et al. 2022) although adjustments can be made setting the appropriate extinction coefficient if the type of DNA or RNA is known. In this regard, another advanced technique fluorometric DNA quantification techniques have an added advantage. In fluorometric techniques, the kits contain an intercalating dye such as either PicoGreen or AccuGreen. These dyes categorically bind to the double stranded DNA. The fluorescence signal is measured and it corelates with the DNA concentration. The Qubit fluorometer and Qubit High Sensitivity quantification kit can be used to measure the samples in the range of 10 pg/μL to 100 ng/μL. The distinguishing between the single and double stranded DNA is another added advantage of this technique (Paul et al. 2021).

The discussed techniques can quantify DNA and to some extent can also be helpful in assessing the quality of the DNA samples. However, this much information is not sufficient enough for the processing of the samples for forensic DNA analysis. All these techniques provide the quantity of whole DNA, which may include the plant DNA, animal DNA, microbial DNA or human DNA. Such exogenous DNA contamination is common in forensic biological samples. However, any source of DNA except of human origin is of no use for a forensic DNA examiner aiming to perform a human identification. In this regard, the quantification of DNA using real-time quantitative PCR is considered to be the gold standard technique. The beauty of multiplexing of this technique is capable of quantifying multiple targets of genomic DNA in a single reaction. The most advanced form of DNA quantification kits can simultaneously quantify the total human DNA concentration, amount of degradation, amount of PCR inhibitors, and the ratio of male and female DNA in a mixed source sample (Lee et al. 2014). The commonly used RT-PCR kits used in various forensic DNA analysis laboratories and their characteristic features is given in Table 5.3.

DNA can be quantified by using qRT-PCR in a real-time manner. However, many workers believe that, this quantification method is not an absolute quantification method. Rather, it is a relative quantification method. In this method, various dilutions of a standard DNA concentration are prepared and run in the instrument. The concentration of the unknown sample is measured with respect to the fluorescence intensity of the standard concentrations. In this regard, some advanced technologies have been discovered to measure the absolute quantity of DNA. One of the emerging techniques is the microfluidic digital PCR (dPCR) technique. This technique allows the absolute quantification of DNA found in very small amounts, i.e., when present in less than 200 copies (Sanders et al. 2011). This technique shows promising applications in the field of absolute quantification of DNA even in the presence of potential PCR inhibitors (Poh et al. 2020).

Table 5.3 Features of the commercially available RT-PCR kits for quantification of DNA

Sl. No.	Name of the kit	Manufacturer	Targets	Output parameters
1	Quantifiler Trio	Thermo Scientific	• Human target (small autosomal) (80 bp)	• Total human DNA
			• Human target (large autosomal) (214 bp)	• M: F ratio
			• Human male target (75 bp)	• Degradation index
			• Internal PCR control (130 bp)	• Presence/absence of PCR inhibitors
2	Quantifiler HP		• Human target (small autosomal) (80 bp)	• Total human DNA
			• Human target (large autosomal) (214 bp)	• Degradation index
			• Internal PCR control (130 bp)	• Presence/absence of PCR inhibitors
3	Quantifiler Duo		• Human target (140 bp)	• Total human DNA
			• Human male target (130 bp)	• M:F ratio
				• Presence/absence of PCR inhibitors
4	Quantifiler® Y human male DNA quantification kit		• Human sex determining region Y gene (SRY) (64 bp)	• Total human male DNA
5	Investigator® Quantiplex HYres Kit	Qiagen	• Human target	• Total human DNA
			• Human male target	• M:F ratio
			• Internal PCR control	• Presence/absence of PCR inhibitors
6	Investigator® Quantiplex Pro Kit		• Human target (small autosomal) (91 bp)	• Total human DNA
			• Human target (large autosomal) (353 bp)	• M:F ratio
			• Human male target (81 bp)	• Degradation index
			• Internal PCR control (434 bp)	• Presence/absence of PCR inhibitors
8	PowerQuant® system	Promega Corporation	• Human target	• Total human DNA
			• Human male	• M:F ratio
			• Degradation	• Degradation index
			• Internal PCR control	• Presence/absence of PCR inhibitors
9	Plexor® HY system		• Human autosomal (99 bp)	• Total human DNA
			• Human Y (133 bp)	• M:F ratio
			• Internal PCR control (150 bp)	• Presence/absence of PCR inhibitors
			• Passive reference	

5.9 DNA Separation Techniques

During the inception of forensic DNA analysis, slab gel electrophoresis was used to separate the amplified DNA fragments. In early 1980s, capillary electrophoresis-based separation of DNA fragments was developed and was found useful in forensic DNA analysis. CE-based separation requires less sample and solvent input and less run time, as well as offering a higher sample throughput and analysis sensitivity (Warner 1994). CE instruments were available commercially by 1988. Some of the nascent models of CE-based systems include the Dionex CESIA CE system (Dionex, Sunnyvale, CA, USA), the Applied Biosystems 270A-HT CE system (Applied Biosystems, Foster City, CA, USA), the Beckman P/ACETM 2050 CE instrument (Beckman Instruments, Fullerton, CA, USA), and the Crystal CE System Model 310 (Unicam Corporation, Madison, WI, USA) (Burgardt and Weissenberger 2022).

Routine forensic DNA analysis involves the separation of amplified fragments by the CE technique. In this step, the amplified fragments are separated based on their size and labeled dye. Subsequently the length-based STR alleles are determined and a comparison is made between the questioned sample and reference sample. Over the years, the CE technique has seen much advancement in terms of their throughput, automation, and resolution capability. Currently, two manufacturers provide sequencers that are widely used in a forensic DNA laboratory. The most commonly used sequencer is the 3500 series from Applied Biosystems. The 3500 and 3500xL series of genetic analyzers have 8 and 24 capillaries, respectively. These instruments use 505 nm solid-state, long-life laser for the detection of the fluorescently labeled amplified fragments. In comparison to the old generation sequencers, 3500 series sequencers have radio frequency identification (RFID) technology that records key consumable administrative information, tracks usage and warns users of replacement status. For a single run, approximately 45 min is required and samples in two 96-welled plates can be loaded in the machine simultaneously (Iyavoo et al. 2017). Recently, another mid-throughput instrument has been launched by Thermo Scientific in the name of the SeqStudio Flex Genetic Analyzers. They are available in both 8 and 24 capillary configurations. Four plate capacity, continuous plate loading, urgent sample reprioritization, and complete scheduling flexibility are few important highlights of these instruments (Aumsuwan et al. 2022).

A significant advancement in the field of capillary electrophoresis took place by the commercial availability of DNA sequencers from another manufacturer, i.e., Promega Corp. The Spectrum Compact CE System by Promega has the capability of single-base resolution, integrated touch screen operation, 6-dye compatible CE analysis and use of class-1 laser product. The high throughput Spectrum CE Systems have four plate capacity, and access to 8-dye technology. This significant advancement in the field of capillary electrophoresis allows to analyze an increased number of genetic markers simultaneously to increase the resolution of a DNA profile (Burgardt and Weissenberger 2022). In this regard, the first 8-Dye PowerPlex® STR Multiplex System is in the market compatible with the CE system which accommodates Amelogenin, 23 autosomal STR loci, 11 Y-STR loci, and two quality indicators in a single multiplex. Some of the 8-dye enabled STR kits are also validated as per the SWAGDM guidelines (Jiang et al. 2021).

5.10 NGS

One of the recent advancements in forensic DNA analysis is the use of the Next Generation Sequencing (NGS) technique. NGS technique provides a high throughput rapid analysis of samples with a higher degree of informativeness. The STR markers routinely used in forensic DNA analysis provides the information on the length-based alleles. When these STR markers are investigated using NGS technique, a significant increase in the number of possible alleles have been observed. This is due to the presence of nucleotide sequence variations within the STR markers. Same length-based allele having different intervening sequence variations is called as isometric heterozygous condition. Such conditions are not a rare phenomenon. A recent study on 31 autosomal STR markers showed the presence of isometric heterozygous conditions in 20 markers in the central Indian population (Dash et al. 2021b). Besides, mutations present in the flanking region of the STRs can also be detected to increase the informativeness of a STR marker using the NGS technique.

Biallelic markers are propounding to be used in future for forensic DNA analysis. As we discussed in the previous chapter, such markers like SNP markers are mostly biallelic in nature and tri or tetra allelic form of such markers are rare. Due to their low allelic variation, SNP markers provide less discrimination among samples. However, the discrimination power can be increased by integrating a large number of SNP markers for a single sample analysis. Analysis of 40–60 SNPs provide the same discrimination level as provided by 13–15 STR markers only (Butler et al. 2007). However, the most commonly used DNA separation techniques are based on 6-dye chemistry. Thus, a huge number of SNP markers cannot be analyzed simultaneously by using the routinely used CE-based analysis. NGS technique can accommodate the analysis of a large number of SNP markers simultaneously. The use of NGS technique is deemed to be of the future of forensic DNA analysis in combination with the SNP markers.

SNP markers have been categorized based on their application, such as, Identity informative (ii) SNPs, lineage-informative (li) SNPs, ancestry-informative (ai) SNPs, phenotype-informative (pi) SNPs, and pharmacogenetic SNPs (Pakstis et al. 2012). These SNP markers can be explored using NGS technique as per the requirement of the investigation process. As a large set of SNP markers are required for the accurate prediction of either lineage or ancestry, they can only be analyzed using NGS technique. Besides the use of STR and SNP markers, mitochondrial DNA analysis is also improved significantly using NGS technique. Sanger sequencing technique is the most efficient for the amplicons ranging from 100 to 800 bp (Crossley et al. 2020). Forensic application of mitochondrial DNA analysis is limited to the sequencing of HVI (positions 16,024–16,365), HVII (positions 73–340), and occasionally HVIII (positions 438–574) (Lutz et al. 2000). The analysis of hypervariable regions shows limited applications regarding the maternal lineage of an individual. Besides, mutations and heteroplasmy complicates the interpretation of mitochondrial DNA analysis process. Now-a-days, whole mitochondrial DNA analysis approach has gained pace in forensic DNA analysis. Since the introduction of NGS technique in 2005, many workers have started sequencing the whole mitochondrial genome to improve the human identification process (Connell et al. 2022). Owing to the large size of mtDNA, i.e., 16,569 bp, Sanger sequencing technique is of limited use. Thus, NGS technique is useful in generating the sequences of the whole mitochondrial genome in small time with accurate sequence information.

5.11 Use of Automation in Forensic DNA Analysis

The routine workflow of forensic DNA analysis is sampling, extraction of DNA, quantification of DNA, amplification of genetic markers, capillary electrophoresis, and interpretation of the DNA profile. The whole process of sample dependent and time consuming as well. As described previously the technology is continuously migrating toward rapid error-free analysis of forensic DNA. Hence, in order to minimize human intervention and generating a quicker result many automated platforms are now available. The most commonly used automated platforms include the automated DNA extraction systems. These automated systems work on the principle of both column-based and magnetic beads-based approach of DNA purification.

The most common automated DNA extraction systems used in forensic DNA analysis laboratories include AutoMate Express™ (Thermo Scientific), EZ1® Advanced XL (Qiagen), and Maxwell® 16 (Promega Corp.). AutoMate Express™ offers two different protocols for the lysis of the samples, the PrepFiler™ lysis buffer or PrepFiler BTA™ lysis buffer. The former is used for the routine forensic samples, whereas, the later is used for the tough samples such as bones, teeth and adhesive samples. This system works on the magnetic bead-based principle and simultaneously extract DNA from 13 samples (Liu et al. 2012). EZ1®

Advanced XL instruments also work on magnetic beads-based technique. It allows the high-quality extraction of genomic DNA and RNA from forensic biological samples (Barbaro et al. 2019). It has a throughput of up to 14 samples per run, though another high throughput platform, i.e., EZ2 is available for simultaneous extraction of DNA from up to 24 samples. Additional feature of the Qiagen DNA extraction systems is the presence of UV lamp to minimize the level of contamination. Promega Corp. offers a wide variety of DNA extraction systems, all of them are based on the magnetic beads-based principle. Maxwell® RSC and Maxwell® CSC systems allow the simultaneous extraction of DNA from 16 samples, whereas, Maxwell® RSC 48 and Maxwell® CSC 48 models allow extraction from 48 samples simultaneously.

Though most of the automated DNA extraction platforms work on the magnetic beads-based principles, a few automated systems are also available working on the column-based principle. One such widely used instrument is the QIAcube (Qiagen). This instrument can simultaneously extract DNA, RNA or protein from 12 samples. Besides the QIAcube (230 V) HID Diff. Washing Stations allow the automation of the differential extraction process. This system is highly useful in processing of mixed biological samples of sexual assault origin, where DNA from sperm fraction and epithelial cell fractions are extracted separately (Goldstein et al. 2020).

In order to avoid manual intervention in the routine forensic DNA workflow process, many automated liquid handling systems are commercially available. Such instruments use disposable plastic wares and carry out the pipetting processing automatically. In many systems machines, thermal cycler, RT-PCR and other instruments can be/has been integrated which facilitates the process. A few features of some of these liquid handling systems are described in Table 5.4. The use of automated liquid handling systems is useful for preparation of dilutions, PCR plate setup, normalization of DNA, post-PCR clean up, NGS library preparation, and many other applications. These systems help in the reducing manual processing of samples, minimizing the risk of contamination, reduced time of pipetting, higher throughput, better reproducibility, stress injuries to humans, and protection from hazardous or infectious samples (Torres-Acosta et al. 2022).

5.12 RAPID DNA Analysis

The traditional forensic DNA analysis is a multistep process which technically requires 10–12 h to complete. Realistically, the overall workflow takes few weeks to months depending on various factors including the laboratory backlogs. To fasten the process, an integrated system performing all the required steps of DNA profiling has been envisioned. This system not only minimizes the time required for DNA analysis, but also offers potential for its application on filed and transportability. The presently available RAPID instruments work on the principle of "sample in, profile out." Since the inception of RAPID DNA technology, two commercially available techniques are present in the market to date. They include ANDE corporation (Waltham, MA) and Thermo Fisher Scientific.

In order to facilitate the use of Rapid DNA technology, the Rapid DNA act was signed into US Federal laws on 18th August, 2017 by amending the DNA Identification Act of 1994. This act allows the generation of DNA profiles of arrestees in police stations using Rapid DNA platforms. Subsequently, an addendum was made in the FBI's Quality Assurance Standards (QAS). The earlier term "Rapid DNA analysis (RDA)' has been replaced with 'modified Rapid DNA analysis (mRDA)." RDA defines the fully automated process of developing a DNA profile without human intervention, whereas, the term "modified Rapid DNA analysis (mRDA)" is defines the fully automated process of developing a DNA profile, but with human interpretation and technical review of the resulting profile.

5.12.1 ANDE System

The first Rapid DNA system was developed in 2012 in the form of the DNAscan™ Rapid DNA Analysis System (GE Healthcare Life Sciences

Table 5.4 Commercially available automated liquid handling systems useful in forensic DNA analysis workflow

Systems	Manufacturer	Features
QIAsymphony	Qiagen	• Purification of DNA
		• Automated transfer of elutes for assay set up
		• Optimization of PCR work flow
IonChef	Thermo Scientific	• Automated library preparation
		• Template preparation
		• Chip loading
Microlab Prep, Microlab NIMBUS, Microlab STAR, Microlab STAR V, Microlab VINTAGE	Hamilton	• Multichannel liquid handler
		• Sample traceability
		• Flexible options for labware transportation
		• Integrated modules for more fully automated workflows
JANUS Mini Automated Liquid Handler, JANUS AJM8001 8-tip Varispan	Perkin Elmer	• Library preparation
		• Automated PCR set up
		• Automated small scale protein purification
		• Automated nucleic acid quantitation and normalization
Biomek FX, Biomek NX, Biomek 3000, Biomek i5, Biomek i7	Beckman Coulter	• Medium to high throughput
		• 0.5–5000 μL pipetting volume ranges
		• Single or dual heads
		• Extraction
		• qPCR/PCR set up
		• NGS library preparation
		• Microarray target preparation
Bravo automated liquid handling platform	Agilent	• Dispensing range dispense from 300 nL to 250 μL
		• Versatile 96 and 384 pipette heads
		• Optional on-deck thermal cycler (ODTC)
epMotion® 5070	Eppendorf	• qPCR setup without cross contamination
		• Faster and more reproducible cell assays
		• Automated NGS library preparation

and NetBio). Later NetBio changed its named as ANDE Corporation. Furthermore, DNAscan™ was also evolved to ANDE 4C System which received NDIS approval by FBI in 2016.

5.12.1.1 DNAscan™ and ANDE 4C System

The BioChipSet Swab, the BioChipSet Cassette, and the DNAscan™/ANDE 4C instrument are three major components of this system. Purification of DNA, amplification of 16 STR loci, and separation of amplified fragments by capillary electrophoresis were carried out in the BioChipSet Cassette. The required consumables and chemicals are preloaded into the cassette. There is an integrated expert system software which analyses the data and subsequently provides the result. Though this system underwent extensive validation, the NDIS approval of this system was retracted later. The retraction of NDIS approval was due to the increase in the number of core-STR loci to 20 from previous 13 CODIS loci.

5.12.1.2 ANDE 6C System

This is an updated version of the ANDE 4C systems to accommodate 20 expanded-CODIS STR markers. ANDE 4C systems received NDIS approval on 1st June, 2018. Three components of this system include the ANDE swab, the A-chip or I-chip, and the instrument. ANDE swab facilitates collection of samples, its drying and prevention of spoilage during storage. The A-chip (abbreviation for arrestee) is used for the reference buccal swab analysis, whereas, the I-chip (abbreviation for Investigative) is used for the challenging samples such as body-fluids, bone, and tissue samples. When I-chip is used, an additional concentration step is performed which maximizes the quantity of DNA. Using the chip, the DNA is purified by silica-based guanidinium salt extraction principle, followed by amplification of 27 STR markers, separation of fluorescently tagged amplified fragments, and detection of fragments using laser-induced fluorescence. Finally, the electropherogram is exported and reviewed by a DNA analyst. The whole analysis process takes around 94 min and 106 min when A and I-chips are used respectively.

5.12.2 Applied Biosystems Systems

5.12.2.1 RapidHIT System

The first generation of RapidHIT systems were developed and validated in 2013 due to their success rate, sensitivity, and tolerance to PCR inhibitors. This is a completely integrated system capable of processing five samples and three control samples simultaneously. RapidHIT system is compatible with both PowerPlex 16HS and GlobalFiler Express kit chemistry. The complete process takes around 90 min for PowerPlex 16HS chemistry and around 120 min for Globalfiler chemistry. This system has a single use kit having a series of individual cartridges, i.e., a sample cartridge, a control cartridge, an anode cartridge with polyacrylamide gel, and a buffer cartridge for CE separation. Up to four sample swabs are processed in the same cartridge. In the control cartridge, one sample swab and three control samples are processed including, a positive control, a negative control, and an allelic ladder.

5.12.2.2 RapidHIT ID System

The RapidHIT system was upgraded to the next generation in the year 2016 in the form of RapidHIT ID system. The upgradation of the system includes an increase in efficiency, increase in capacity, reduction of footprint, and less user intervention. Another significant change made in the system is the replacement of the previous cartridge having four subsystem configurations with a single cartridge for each sample. Another substantial advancement in the upgraded system includes the development of the new RapidLINK software for data analysis and interpretation of DNA profile.

5.13 Conclusion

In the last two decades, the forensic DNA analysis process has seen significant advancements in the form of its sensitivity, reliability, robustness, and rapidity. The world has witnessed the migration of 13 CODIS core loci to 20 expanded-CODIS loci. With this increase in the number of STR markers, the commercially available kits have also been modified accordingly. The high throughput instruments are available in the market to take of the increasing demand of the forensic DNA analysis work load. Still, a huge burden of pending cases in different laboratories globally poses a huge challenge to the techniques used currently. Rapid DNA technology have been developed to meet this need of the forensic science laboratories. Despite many advancements in the field of rapid DNA analysis, the scope of development in future includes smaller and faster instruments, increased cost-effectiveness, reduced handling time, and updated policy and legislation framework globally.

References

Alaeddini R (2012) Forensic implications of PCR inhibition—a review. Forensic Sci Int Genet 6:297–305. https://doi.org/10.1016/j.fsigen.2011.08.006

Andersen MM, Mogensen HS, Eriksen PS, Morling N (2017) Yfiler® plus population samples and dilution series: stutters, analytic thresholds, and drop-out probabilities. Int J Legal Med 131:1503–1511. https://doi.org/10.1007/s00414-017-1568-8

Anderst J (2011) The forensic evidence kit. In: Jenny C (ed) Child abuse and neglect: diagnosis, treatment, and evidence. Saunders/Elsevier, St. Louis, MO, pp 106–111. https://www.clinicalkey.com/#!/browse/book/3-s2.0-C20090389746

Anslinger K, Bayer B, Mack B, Eisenmenger W (2007) Sex-specific fluorescent labelling of cells for laser microdissection and DNA profiling. Int J Legal Med 121:54–56. https://doi.org/10.1007/s00414-005-0065-7

Aumsuwan PK, Berosik S, Chege P, Sannicandro A, Samsani S, Bodeau J, Foster G, Rodriguez D, Lagier E, Chee S, Nelson S, Bornarth C (2022) Genetic analysis instrumentation innovations built into the SeqStudio™ flex genetic analyzers. Forensic Sci Int Genet Suppl Ser 8:228–231. https://doi.org/10.1016/j.fsigss.2022.10.044

Barbaro A, Cormaci P, Marca AL (2019) DNA extraction from soil by EZ1 advanced XL (Qiagen). Forensic Sci Int 299:161–167. https://doi.org/10.1016/j.forsciint.2019.04.004

Bordelon H, Russ PK, Wright DW, Haselton FR (2013) A magnetic bead-based method for concentrating DNA from human urine for downstream detection. PLoS One 8:e68369. https://doi.org/10.1371/journal.pone.0068369

Bruijns B, Hoekema T, Oomens L, Tiggelaar R, Gardeniers H (2022) Performance of spectrophotometric and fluorometric DNA quantification methods. Analytica 3(3):371–384. https://doi.org/10.3390/analytica3030025

Budowle B, Klevan L, Eisenberg AJ (1994) RFLP typing: a new highly polymorphic VNTR locus and chemiluminescent detection. In: Bär W, Fiori A, Rossi U (eds) Advances in forensic haemogenetics, vol 5. Springer, Berlin. https://doi.org/10.1007/978-3-642-78782-9_62

Burgardt N, Weissenberger M (2022) First experiences with the spectrum compact CE system. Int J Legal Med 136:55–72. https://doi.org/10.1007/s00414-021-02673-1

Butler JM, Coble MD, Vallone PM (2007) STRs vs. SNPs: thoughts on the future of forensic DNA testing. Forensic Sci Med Pathol 3:200–205

Castella V, Dimo-Simonin N, Brandt-Casadevall C, Mangin P (2004) Consensus profiles and databasing of casework samples amplified with 34 PCR cycles: an empirical approach. Int Congr Ser 1261:532–534. https://doi.org/10.1016/S0531-5131(03)01640-6

Castella V, Dimo-Simonin N, Morerod ML, Mangin P (2006) In-house validation of the PCR amplification kit TMentypeR ArgusX-ULr. Int Congr Ser 1288:310–312. https://doi.org/10.1016/j.ics.2005.08.042

Chen J, Kobilinsky L, Wolosin D, Shaler R, Baum H (1998) A physical method for separating spermatozoa from epithelial cells in sexual assault evidence. J Forensic Sci 43:114–118

Comey CT, Budowle B (1991) Validation studies on the analysis of the HLA DQ alpha locus using the polymerase chain reaction. J Forensic Sci 36:1633–1648

Connell JR, Benton MC, Lea RA, Sutherland HG, Haupt LM, Wright KM, Griffiths LR (2022) Evaluating the suitability of current mitochondrial DNA interpretation guidelines for multigenerational whole mitochondrial genome comparisons. J Forensic Sci 67:1766–1775. https://doi.org/10.1111/1556-4029.15097

Cornelis S, Fauvart M, Gansemans Y, Plaetsen ASV, Colle F, Wiederkehr RS, Deforce D, Stakenborg T, Nieuwerburgh FV (2018) Multiplex STR amplification sensitivity in a silicon microchip. Sci Rep 8:9853. https://doi.org/10.1038/s41598-018-28229-9

Crossley BM, Bai J, Glaser A, Maes R, Porter E, Killian ML, Clement T, Toohey-Kurth K (2020) Guidelines for sanger sequencing and molecular assay monitoring. J Vet Diagn Investig 32:767–775. https://doi.org/10.1177/1040638720905833

Dash HR, Shrivastava P, Das S (2020) Principles and practices of DNA analysis: A laboratory manual for forensic DNA typing, Springer protocols handbooks. Springer, New York. https://doi.org/10.1007/978-1-0716-0274-4

Dash HR, Kaitholia K, Kumawat RK, Singh AK, Shrivastava P, Chaubey G, Das S (2021a) Sequence variations, flanking region mutations, and allele frequency at 31 autosomal STRs in the central Indian population by next generation sequencing (NGS). Sci Rep 11:23238. https://doi.org/10.1038/s41598-021-02690-5

Dash HR, Rawat N, Vajpayee K, Shrivastava P, Das S (2021b) Useful autosomal STR marker sets for forensic and paternity applications in the central Indian population. Ann Hum Biol 48:37–48. https://doi.org/10.1080/03014460.2021.1877353

Elliott K, Hill DS, Lambert C, Burroughes TR, Gill P (2003) Use of laser microdissection greatly improves the recovery of DNA from sperm on microscope slides. Forensic Sci Int 137:28–36. https://doi.org/10.1016/s0379-0738(03)00267-6

Elwick K, Mayes C, Hughes-Stamm S (2018) Comparative sensitivity and inhibitor tolerance of GlobalFiler® PCR amplification and investigator® 24plex QS kits for challenging samples. Legal Med 32:31–36. https://doi.org/10.1016/j.legalmed.2018.01.006

Ensenberger MG, Lenz KA, Matthies LK, Hadinoto GM, Schienman JE, Przech AJ, Morganti MW, Renstrom DT, Baker VM, Gawrys CM, Hoogendoorn M, Steffen CR, Martín P, Alonso A, Olson HR, Sprecher CJ, Storts DR (2016) Developmental validation of the

PowerPlex® fusion 6C system. Forensic Sci Int Genet 21:134–144. https://doi.org/10.1016/j.fsigen.2015.12.011

Ferreira-Silva B, Fonseca-Cardoso M, Porto MJ, Magalhães T, Cainé L (2018) A comparison among three multiplex Y-STR profiling kits for sexual assault cases. J Forensic Sci 63:1836–1840. https://doi.org/10.1111/1556-4029.13757

Garvin AX, Holzinger R, Berner F, Krebs W, Hostettler B, Lardi E, Hertli C, Quartermaine R, Stamm C (2013) The forensiX evidence collection tube and its impact on DNA preservation and recovery. Biomed Res Int 2013:105797. https://doi.org/10.1155/2013/105797

Gehrig C, Castella V (2013) Internal validation study of the PowerPlex® Y23 system. Forensic Sci Int Genet Suppl Ser 4:e79–e80. https://doi.org/10.1016/j.fsigss.2013.10.040

Gill P, Jeffreys AJ, Werrett DJ (1985) Forensic application of DNA 'fingerprints'. Nature 318:577–579

Goldstein MC, Cox JO, Seman LB, Cruz TW (2020) Improved resolution of mixed STR profiles using a fully automated differential cell lysis/DNA extraction method. Forensic Sci Res 5:106–112. https://doi.org/10.1080/20961790.2019.1646479

Golubeva Y, Salcedo R, Mueller C, Liotta LA, Espina V (2013) Laser capture microdissection for protein and NanoString RNA analysis. Methods Mol Biol 931:213–257. https://doi.org/10.1007/978-1-62703-056-4_12

Hares DR (2015) Selection and implementation of expanded CODIS core loci in the United States. Forensic Sci Int Genet 17:33–34. https://doi.org/10.1016/j.fsigen.2015.03.006

Horsman KM, Barker SLR, Ferrance JP, Forrest KA, Koen KA, Landers JP (2005) Separation of sperm and epithelial cells in a microfabricated device: potential application to forensic analysis of sexual assault evidence. Anal Chem 77:742–749. https://doi.org/10.1021/ac0486239

Hu Q, Liu Y, Yi S, Huang D (2015) A comparison of four methods for PCR inhibitor removal. Forensic Sci Int Genet 16:94–97. https://doi.org/10.1016/j.fsigen.2014.12.001

Huang S, Jin X, Zhang H, Jin H, Ren Z, Wang Q, Liu Y, Ji J, Yang M, Zhang H, Zheng X, Song D, Zheng B, Huang J (2022) Developmental validation of the novel five-dye-labeled multiplex autosomal STR panel and its forensic efficiency evaluation. Front Genet 13:897650. https://doi.org/10.3389/fgene.2022.897650

Iyavoo S, Draus-Barini J, Haizel T (2017) Validation of ABI 3500xL genetic analyzer after decommissioned and recommissioned at new premises. Forensic Sci Int Genet Suppl Ser 6:e542–e543. https://doi.org/10.1016/j.fsigss.2017.09.197

Jiang B, Qu W, Wang F, Zhang L, Rong H, Li J, Wen D, Zeye MMJ, He W, Wang C, Xing H, Zhang T, Jin C, Chen L, Liu Y, Cai J, Zha L (2021) Development and validation of novel 8-dye short tandem repeat multiplex system for forensic applications. Int J Legal Med 135:2263–2274. https://doi.org/10.1007/s00414-021-02695-9

Kaunitz JD (2015) The discovery of PCR: ProCuRement of divine power. Dig Dis Sci 60:2230–2231. https://doi.org/10.1007/s10620-015-3747-0

Kawane K, Motani K, Nagata S (2014) DNA degradation and its defects. Cold Spring Harb Perspect Biol 6:a016394. https://doi.org/10.1101/cshperspect.a016394

Kloosterman AD, Budowle B, Daselaar P (1993) PCR-amplification and detection of the human D1S80 VNTR locus. Amplification conditions, population genetics and application in forensic analysis. Int J Legal Med 105:257–264. https://doi.org/10.1007/BF01370382

Lee SB, McCord B, Buel E (2014) Advances in forensic DNA quantification: a review. Electrophoresis 35:3044–3052. https://doi.org/10.1002/elps.201400187

Li C, Han J, Ren W, Ji A, Xu X, Hu L (2011) DNA profiling of spermatozoa by laser capture microdissection and low volume-PCR. PLoS One 6(8):e22316. https://doi.org/10.1371/journal.pone.0022316

Lin SW, Li C, Ip SCY (2017) A selection guide for the new generation 6-dye DNA profiling systems. Forensic Sci Int Genet 30:34–42. https://doi.org/10.1016/j.fsigen.2017.05.010

Liu JY, Zhong C, Holt A, Lagace R, Harrold M, Dixon AB, Brevnov MG, Shewale JG, Hennessy LK (2012) AutoMate express™ forensic DNA extraction system for the extraction of genomic DNA from biological samples. J Forensic Sci 57:1022–1030. https://doi.org/10.1111/j.1556-4029.2012.02084.x

Liu W, Chen W, Liu R, Ou Y, Liu H, Xie L, Lu Y, Li C, Li B, Cheng J (2015) Separation of sperm and epithelial cells based on the hydrodynamic effect for forensic analysis. Biomicrofluidics 9:044127. https://doi.org/10.1063/1.4928453

Lutz S, Wittig H, Weisser HJ, Heizmann J, Junge A, Dimo-Simonin N, Parson W, Edelmann J, Anslinger K, Jung S, Augustin C (2000) Is it possible to differentiate mtDNA by means of HVIII in samples that cannot be distinguished by sequencing the HVI and HVII regions? Forensic Sci Int 113:97–101. https://doi.org/10.1016/s0379-0738(00)00222-x

Magalhães T, Dinis-Oliveira RJ, Silva B, Corte-Real F, Vieira DN (2015) Biological evidence management for DNA analysis in cases of sexual assault. Sci World J 2015:365674. https://doi.org/10.1155/2015/365674

McAlister C (2011) The use of fluorescence in situ hybridisation and laser microdissection to identify and isolate male cells in an azoospermic sexual assault case. Forensic Sci Int Genet 5:69–73. https://doi.org/10.1016/j.fsigen.2010.04.008

Micka KA, Sprecher CJ, Lins AM, Theisen Comey C, Koons BW, Crouse C, Endean D, Pirelli K, Lee SB, Duda N, Ma M, Schumm JW (1996) Validation of multiplex polymorphic STR amplification sets developed for personal identification applications. J Forensic Sci 41:582–590

Miller DN, Bryant JE, Madsen EL, Ghiorse WC (1999) Evaluation and optimization of DNA extraction and purification procedures for soil and sediment samples. Appl Environ Microbiol 65:4715–4724. https://doi.org/10.1128/aem.65.11.4715-4724.1999

Miyazaki T, Hara M, Ichiki A, Yamamoto Y, Takada A, Kido A, Nodera M, Yanagisawa H, Suzuki H, Saito K (2008) An efficient novel method for analyzing STR loci from a single sperm captured by laser microdissection. Forensic Sci Int Genet Suppl Ser 1:437–438. https://doi.org/10.1016/j.fsigss.2007.10.074

Mulero JJ, Chang CW, Lagacé RE, Wang DY, Bas JL, McMahon TP, Hennessy LK (2008) Development and validation of the AmpFlSTR MiniFiler PCR amplification kit: a MiniSTR multiplex for the analysis of degraded and/or PCR inhibited DNA. J Forensic Sci 53:838–852. https://doi.org/10.1111/j.1556-4029.2008.00760.x

Norris JV, Evander M, Horsman-Hall KM, Nilsson J, Laurell T, Landers JP (2009) Acoustic differential extraction for forensic analysis of sexual assault evidence. Anal Chem 81:6089–6095. https://doi.org/10.1021/ac900439b

Pakstis AJ, Fang R, Furtado MR, Kidd JR, Kidd KK (2012) Mini-haplotypes as lineage informative SNPs and ancestry inference SNPs. Eur J Hum Genet 20:1148–1154. https://doi.org/10.1038/ejhg.2012.69

Paul P, Rajput S, Joshi P, Naithani M, Chowdhury N, Rao S, Pai MO (2021) Comparison of Fluorometric and UV spectrophotometric findings for DNA isolated from formalin-fixed paraffin-embedded blocks, fine needle aspiration cytology smears, and blood. Cureus 13:e19583. https://doi.org/10.7759/cureus.19583

Poh TY, Ali NABM, Chan LLY, Tiew PY, Chotirmall SH (2020) Evaluation of droplet digital polymerase chain reaction (ddPCR) for the absolute quantification of *Aspergillus* species in the human airway. Int J Mol Sci 21:3043. https://doi.org/10.3390/ijms21093043

Sanders R, Huggett JF, Bushell CA, Cowen S, Scott DJ, Foy CA (2011) Evaluation of digital PCR for absolute DNA quantification. Anal Chem 83:6474–6484. https://doi.org/10.1021/ac103230c

Scherer M, König M, Bussmann M, Prochnow A, Peist R (2015) Development and validation of the new Investigator1 Argus X-12 QS kit. Forensic Sci Int Genet Suppl Ser 5:e256–e257. https://doi.org/10.1016/j.fsigss.2015.09.102

Sidstedt M, Hedman J, Romsos EL, Waitara L, Wadsö L, Steffen CR, Vallone PM, Rådström P (2018) Inhibition mechanisms of hemoglobin, immunoglobulin G, and whole blood in digital and real-time PCR. Anal Bioanal Chem 410:2569–2583. https://doi.org/10.1007/s00216-018-0931-z

Siegel JA, Saukko PJ, Houck MM (2013) Encyclopedia of forensic sciences, 2nd edn. Elsevier, London

Torres-Acosta MA, Lye GJ, Dikicioglu D (2022) Automated liquid-handling operations for robust, resilient, and efficient bio-based laboratory practices. Biochem Eng J 188:108713. https://doi.org/10.1016/j.bej.2022.108713

Tozzo P, Ponzano E, Spigarolo G, Nespeca P, Caenazzo L (2018) Collecting sexual assault history and forensic evidence from adult women in the emergency department: a retrospective study. BMC Health Serv Res 18:383. https://doi.org/10.1186/s12913-018-3205-8

Vraneš M, Scherer M, Cornelius S, Willuweit M, Kraemer M, Terramagra D, Kohns A, Prochnow A (2022) Developmental validation of the investigator Argus Y-28 QS kit. Forensic Sci Int Genet Suppl Ser 8:274–275. https://doi.org/10.1016/j.fsigss.2022.10.060

Wang DY, Chang CW, Lagacé RE, Calandro LM, Hennessy LK (2012) Developmental validation of the AmpFℓSTR® Identifiler® plus PCR amplification kit: an established multiplex assay with improved performance. J Forensic Sci 57:453–465. https://doi.org/10.1111/j.1556-4029.2011.01963.x

Warner M (1994) Migrating toward capillary electrophoresis. Anal Chem 66(22):1137A–1141A. https://doi.org/10.1021/ac00094a002

White J, Hughes-Stamm S, Gangitano D (2015) Development and validation of a rapid PCR method for the PowerPlex® 16 HS system for forensic DNA identification. Int J Legal Med 129:715–723. https://doi.org/10.1007/s00414-014-1102-1

6 Advancements in Non-human Forensic DNA Analysis

6.1 Introduction

Non-human forensics includes DNA analysis of wildlife, insects, biological material recovered for geolocation or linking suspects or perpetrators to crime scenes, food products for authenticity, and biological agents used to perpetrate crimes such as plants and pathogens. Wildlife forensics encompasses non-human DNA analysis including identification of endangered and legally protected species and analysis of materials left by animal perpetrators or witnesses to crimes. Non-human forensics includes investigations of bacterial and viral agents used in bioterrorism, legal high plants or fungal species used as drugs, pollen transferred to a suspect while committing a crime, and identifying insect colonizers used to predict post-mortem interval (PMI).

Crime scene investigators obtain many samples and may have few clues as to which species' DNA is contained in a sample. Samples may include non-human biological material such as hair, fibers, body fluids or feces found on clothing, bags and purses, or in a car, or on a rug, couch, sheets, toys, tires, or even the undercarriage of a vehicle at a hit and run accident. Vacuum filters can also be a source of microbial DNA and pet dander, hair, and fiber evidence. Biological material recovered from the victim or at a crime scene can be analyzed to determine the identity of the species in a sample.

Methods for determining species include mitochondrial DNA sequencing, quantitative real-time PCR, qualitative PCR, post-PCR melt analysis, gel electrophoresis, fragment analysis, single nucleotide polymorphism (SNP) typing, targeted DNA sequencing, and whole genome sequencing. In this chapter, we will cover frequently encountered non-human DNA contributors identified in forensic evidence and advancements in DNA analysis.

6.2 Animals

Animal contributors to crime scenes may include pets, wildlife, and livestock. Animals can be victims of crimes, perpetrators, or even witnesses. Application of forensic DNA techniques to non-human DNA circumstances in wildlife forensics evidence submitted as part of a criminal case often seeks to answer the question of what species is the contributor (Tobe and Linacre 2010). Animal species are also frequently investigated in forensic cases especially those suspected of being illegally poached and in cases of mislabeled food products such as meat and dairy. Samples may include biological material such as body fluids, hair or fur, flesh, bones and teeth (Eichmann et al. 2004; Müller et al. 2008; Sundqvist et al. 2008; Sato et al. 2010).

According to the American Pet Product Association National Pet Owners Survey, in 2023, approximately 86.9 million U.S. families—or 66% of U.S. households (131.2 million households according to the U.S. Census Bureau

H. R. Dash et al., *Advancements in Forensic DNA Analysis*,
https://doi.org/10.1007/978-981-99-6195-5_6

in 2022)—own a pet (https://www.americanpetproducts.org/press_industrytrends.asp). Pets are frequently found at crime scenes where they may have been perpetrators, victims, or witnesses. Microscopy is used to differentiate and identify the species origin of hairs, fur, and bones from animals by morphology. The color and features of the hair or fiber can be analyzed to classify its source by species (Deedrick and Koch 2004). However, in some cases such as pet bites or attacks, body fluids including saliva or blood (or even urine or feces) may be the submitted evidence (Eichmann et al. 2004; Clarke and Vandenberg 2010). Immunological serological antibody-antigen assays (e.g., ABA HemaTrace cards) can be used to determine if a body fluid is human or not prior to DNA typing. If the source is human, a human DNA typing kit can be used. If the source is not human, another test must be employed to determine the species so that an appropriate STR DNA typing kit or test can be selected to individualize the source (Eichmann et al. 2004; Butler 2005, Müller et al. 2008).

DNA barcoding and PCR assays can be used to identify the species while individual animals can be linked to the genetic profiles obtained from crime scene evidence using short tandem repeat (STR) DNA typing. In DNA barcoding, the highly conserved including 16S rRNA, 18S rRNA, 28S rRNA, cytochrome b (cyt b), beta-actin, cytochrome oxidase I (COI) genes, and mitochondrial highly variable (HV) regions are targeted for sequencing or amplification and detection (Parson et al. 2000; Bellis et al. 2003; Bravi et al. 2004; Caine et al. 2006; Nakaki et al. 2007). DNA barcoding sequences from evidence can be compared to a gene sequence in a database. Animal species differentiation using PCR followed by restriction fragment length polymorphism (RFLP) has been used in domestic animal identification (Bravi et al. 2004). Real-time PCR has been used to develop assays to differentiate and identify common household pets. Unlike end-point PCR, real-time PCR assays do not require post-PCR gel analysis. Using DNA gene sequences available in NCBI GenBank, primers can be designed to target a species of interest; theoretical specificity can be determined using

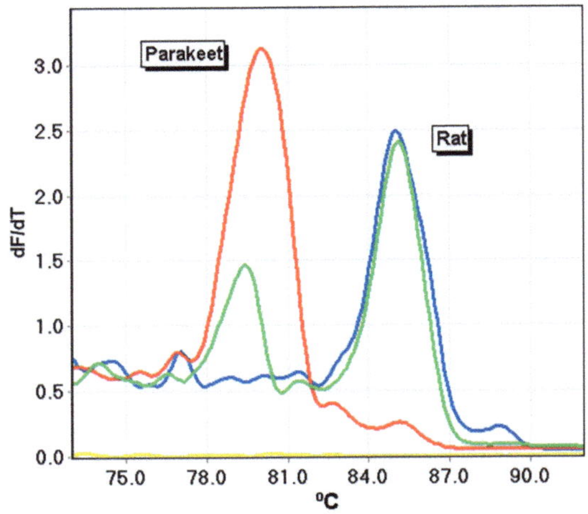

Fig. 6.1 Real-time PCR melt assay for parakeet (red) and rat (blue); the green trace shows a mixed DNA sample of the two pet species with melts at both diagnostic temperatures and the yellow line is the no template control (NTC)

NCBI Blast. A qPCR assay was developed for identifying human, cat, and dog species (Kanthaswamy et al. 2012). PCR high resolution melt (HRM) assays (Fig. 6.1) have been developed for pets including parakeet, rabbit, guinea pig, dog, rat, and hamster (Johnson and Elkins, unpublished results). Two-step direct PCR was used to analyze animal DNA (Kitpipit et al. 2014). FINS (Forensically Informative Nucleotide Sequencing) was developed to identify the animal origin of biological samples (Bartlett and Davidson 1992). A single base extension assay was developed to detect and identify dogs, cats, and humans (Nakaki et al. 2007). Using single hairs with and without roots, a 402-bp segment of mtDNA was used to classify cats into mitotypes (Tarditi et al. 2011). The Meowplex STR assay was developed to identify cats (Butler 2005) and a STR DNA assay was developed for dogs (Eichmann et al. 2004).

6.3 Insects

Forensic entomology is the scientific study of the colonization of a corpse by insects such as flies and beetles and collection of evidence to aid in investigations; a recent review of the field was published by Lutz et al. (2021). Necrophagous insects including blowflies and dermestid beetles can have evidentiary value in forensic cases (Rivers and Geiman 2017). Necrophagous insects are species that feed on and interact with corpses and body fluids within certain crime scene environments (Greenberg 1991). Identification of the species of flies and beetles colonizing bodies after death can aid in determination of the post-mortem interval, chain-of-evidence disputes, and if a corpse was moved (Campobasso et al. 2005; Amendt et al. 2011). Flies can deposit regurgitate and defecate and also leave tarsal tracks which are referred to as fly artifacts (Rivers and Geiman 2017). The deposits and tracks can resemble bloodstain patterns and can be detected by an alternate light source and identified by attenuated total reflectance (ATR) Fourier Transform Infrared (FT-IR) spectroscopy (Bowser et al. 2023). The insects and the fly artifacts can be sampled and analyzed to determine the species by an entomologist or forensic DNA scientist. The DNA can be used to identify the species of the insect colonizer; DNA from the crop also has been recovered for human DNA analysis and to determine the larvae's last meal. Recovering human DNA from fly maggots and artifacts can aid in investigations if the source of the DNA can be identified using a database, suspect, or reference sample. Over 20 years ago, human and insect mitochondrial DNA was recovered from maggots and analyzed (Wells et al. 2001; Linville et al. 2004). STR typing was also performed from maggot crops (Linville et al. 2004). STR typing of human DNA recovered from necrophagous fly larvae that fed on human corpses has been performed; the animal species was identified using the cytB gene (Zehner et al. 2004). Incomplete and complete STR profiles were obtained and mitochondrial HV region typing was also performed (Zehner et al. 2004). Mitochondrial DNA and human STR analyses of blowfly *Aldrichina grahami* maggots' crop contents in a case from central-southern China disavowed that the insect DNA was a contaminant to the case (Li et al. 2011). A study of insect artifacts yielded full STR profiles that matched the source of the DNA (Durdle et al. 2013). A Y-STR profile of human DNA was obtained from *Chrysomya albiceps* larvae fed huma semen (Chamoun et al. 2019). DNA barcoding has been used to analyze insect fragments recovered with soil (Meiklejohn et al. 2018) and genome and transcriptome analysis has been performed for the flesh fly *Sarcophaga bullata* (Martinson et al. 2019). The drawback of human DNA identification from insect larvae is that the quantity of human DNA decreases over time since the larvae fed on the corpse and partial profiles are common (Njau et al. 2016).

6.4 Plants

DNA analysis of plant evidence for crime scene analysis includes identifying commercial plant products that may be mislabeled, samples for geolocation, plant sources of toxins, and investigations of illicit and legal high drugs (Zaya and Ashley 2012). A comparison of DNA extraction methods demonstrated that several methods including the MicroGEM PDQeX phytoGEM system, Omega Bio-tek E.Z.N.A. Plant DS Mini kit, and Qiagen DNeasy Plant Kit can be used to extract DNA from plant species (Ryan et al. 2020).

Controlled substances including morphine codeine are extracted from opium poppy (Elkins 2019). Heroin is a semisynthetic illicit drug synthesized from morphine. Legal high plant drugs including opium poppy seeds, jimson weed, and kratom are not controlled and may be encountered in cases of unexpected deaths (Elkins et al. 2017). Unexpected deaths are those that occur in apparently healthy individuals and are not preceded by significant symptoms or known illness or disease. "Legal high" drugs of abuse are included in the classification of novel or new psychoactive substances (NPS) (Elkins 2019). Legal high plants produce endogenous substances that produce a "high" or altered state and may lead to unexpected and additional physiological responses especially when taken with other substances. As a result, legal highs may pose more of a risk than other drugs of abuse as users have no dosage guidelines and healthcare workers may have little knowledge of these substances. Legal highs attract teens and new drug users seeking a drug-like effect since the substances are widely-accessible and not controlled. Forensic scientists and medical professionals need tools to identify legal highs, especially those from plant and mushroom sources are there few forensic botanists make identifications using morphology. Identification of plant drugs enables health providers to treat patients, enables health departments to raise awareness, and reduces risks to communities. Furthermore, drug paraphernalia with only burned or trace residue of ground plant material are difficult to impossible to identify morphologically. Sanger and next generation DNA sequencing, DNA barcoding, SNP typing, and PCR assays using targeted amplification have been used identify the species contributors found on drug paraphernalia evidence (Aly and Sabri 2015; Elkins et al. 2017; Meiklejohn et al. 2018; Antil et al. 2023). Eukaryotic 18S rRNA, trnL intron, psbA-trnH, rpoC1, rpoB, atpF-atpH, psbK-psbI, cytochrome oxidase subunit I (COI), ribulose 1,5-biphosphate carboxylase (rbcL), maturase K (matK), and internal transcribed spacer I (ITS1) genes sequences are used for species identification in plants (Young et al. 2014; Figueroa et al. 2017; Antil et al. 2023). SNPs have been used to differentiate strains of cannabis (Henry et al. 2020). PCR high resolution melt assays developed by targeting other plant genes have been developed to detect and identify legal high plants DNA from poppy, jimson weed, Hawaiian woodrose, morning glory, wormwood, and kratom (Elkins et al. 2017; Kiesel and Elkins 2019; Cowan and Elkins 2020). Figure 6.2 shows sample PCR HRM data for three "legal high" plants *Ipomoea purpurea* (IP), *Datura stramonium* (DS), and *Cannabis sativa* (CS). DNA methods including PCR HRM have also been used to analyze hallucinogenic fungi (Nugent and Saville 2004; Figueroa et al. 2017; Cowan and Elkins 2018).

Ricin (*Ricinus communis*) has been used in terror attacks (Elkins 2019) and has also been the cause of death in suicide case (Coopman et al. 2009) and an accidental poisoning case (Despott and Cachia 2004). Whereas ricin or castor oil plants are cultivated for castor oil production, the water-soluble ricin toxin can be isolated from the remaining mash by acidic extraction and precipitated to a white powder (Kikkawa et al. 2015; Elkins 2019). The DNA from ricin and abrin plants has been identified using quantitative real-time PCR (Felder et al. 2012). Ricin can be differentiated by DNA sequencing (Chan et al. 2010).

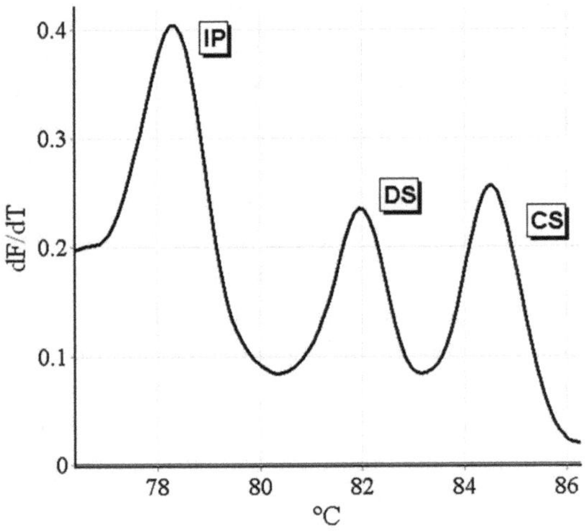

Fig. 6.2 Real-time PCR HRM assay data for three plants morning glory (IP), jimson weed (DS), and marijuana (CS)

Plant-derived material such as flour can be used as a cutting agent in street drugs. White wheat (*Triticum aestivum* L.) flour has been detected and identified in trace evidence samples by partial sequencing of *rbc*L gene and STR analysis with specific cultivars exhibited different patterns (Kikkawa et al. 2015). Plant material and pollen has also been used for geolocation. Geolocation is the process of determining the geographical location of a crime scene or place a victim or suspect has been in contact with. Botanists can differentiate seeds and pollen by morphology using microscopy including scanning electron microscopy (SEM). Alternatively, DNA analysis can be used to determine the species (Elkins and Marlette 2023). In a 1992 case, seed pods from a Palo Verde tree linked a suspect's truck to the crime scene (Yoon 1993).

Commercial wood products are prone to counterfeiting and are often detected as mislabeled using microscopy, pattern characteristics, as well as DNA typing methods (Wiedenhoeft 2016; Knorr-Evans and Parker-Forney 2019; Irwin 2019; Elkins and Marlette 2023). Craft et al. (2007) and Lemes et al. (2002) used microsatellites plant markers to identify oak and mahogany species, respectively. Eighty SNPs were used to identify seven white DNA species (Hipp et al. 2019) and SNPs were also used to identify mahogany (Pakull et al. 2020). The 18S rRNA and isoprene synthase (IspS) genes has been used to identify oak (*Quercus palustris*) (Barta et al. 2017). DNA barcoding using the trnL-trnF and matK DNA sequences were used to identify agarwood species (Tanaka and Ito 2020). DNA barcoding has also been used to identify evergreen species (Nithaniyal et al. 2014). Sequencing of the oak genome can also be used in its identification (Tang et al. 2018). A real-time PCR assay has been applied to agarwood (Mohamed et al. 2012).

6.5 Pathogens

Several pathogens including anthrax, *E. coli*, and *Shigella* have been used in bioterror cases (Murch 2005; Kolavic et al. 1997; Török et al. 1997). Several methods have been developed and tested for the detection and analysis of pathogens and there are several reviews on the subject (Dwivedi and Jaykus 2011; Druml and Cichna-Markl 2014; Law et al. 2015; Switai et al. 2015; Priyanka et al. 2015; Alahi and Mukhopadhyay 2017; Rajapaksha et al. 2018). Quantitative PCR and targeted sequencing assays are highly sensitive and specific and are used for detecting and identifying foodborne and bioterror pathogens (Elkins and Bender 2020). While the gold standard remains to culture pathogens for identification, not all bacteria can be cultured, the process is time-consuming (3–7 days) and labor intensive, and phenotypically similar species cannot be differentiated (Elkins 2019). DNA analysis methods can be automated for surveillance and detect bacterial, viral, and toxin-related genes for a range of pathogens. They can be high throughput and highly sensitive at low cost. The 16S rRNA gene is frequently probed for bacterial species identification (Young et al. 2014) but rpoB, cpn60, tuf, RIF, and gnd genes have also been used (Antil et al. 2023). Some bacteria are mutualistic while others are pathogenic. According to the U.S. Centers for Disease Control, pathogens that frequently cause foodborne illness in the United States include bacteria *Escherichia coli*, *Salmonella enterica*, *Shigella flexneri*, *Listeria monocytogenes*, *Vibrio parahaemolyticus*, *Camplobacter jejuni*, and *Clostridium difficile* as well as viruses such as Norovirus. Consumers may unknowingly ingest the pathogens upon eating uncooked or undercooked meats, shellfish, raw unwashed fruits or vegetables, and/or salads prepared on contaminated surfaces or with contaminated water. Drinking contaminated water can also lead to illness. The CDC estimates that 30% of the population in industrialized countries suffers from illness caused by foodborne pathogens from contaminated food and drinking water. Foodborne pathogens are a major health burden and cause a morbidity and deaths each year. Companies are expected to notify the Food and Drug Administration (FDA) when they test and determine their products have a "reasonable probability" of causing serious adverse health consequences and immediately perform a corrective action. If the sources of the illnesses are caught relatively quickly, they may not be too widespread, if the agent is not transmissible. In a recent case of Norovirus, a transmissible virus, more than 200 people were sickened at a restaurant where food was contaminated with the virus; the public and public health officials were only informed after the restaurant had closed and cleaned. Not informing public health officials led to criminal investigations by the Justice Department (https://wjla.com/news/nation-world/lawsuit-chipotle-covered-up-evidence-of-norovirus-in-tainted-food). In other cases involving *Shigella* and *E. coli*, perpetrators used the pathogens to intentionally sicken individuals (Kolavic et al. 1997; Török et al. 1997). Rapid analysis is important to identify the index case, reduce risks to consumers, prevent outbreaks, identify the presence of a pathogen in air, food or water, and reduce labor costs and time (Meselson et al. 1994). 16S rRNA gene sequencing, whole genome sequencing, and targeted PCR assays are can be used to detect and identify the pathogens. High resolution melt PCR assays are popular for detecting and identifying microorganisms (Kagkli et al. 2012; Gopaul et al. 2014; Elkins et al. 2016; Hoseinpour et al. 2017; Harrison and Hanson 2017; Tamburro and Ripabelli 2017; Liu et al. 2018; Bender et al. 2020; Reese and Elkins 2022). Additional DNA analysis methods include DNA microarrays and loop mediated isothermal amplification (LAMP) assays. Several assays for foodborne pathogens have been developed and reported (Pavlovic et al. 2011; Kagkli et al. 2012; Gopaul et al. 2014; Elkins et al. 2016; Hoseinpour et al. 2017; Harrison and Hanson 2017; Tamburro and Ripabelli 2017; Liu et al. 2018; Bender et al. 2020; Reese and Elkins 2022). Bioterror agents including anthrax have been identified by culture, rapid PCR assays, and DNA sequencing (Pavlovic et al. 2011; Rasko et al. 2011). These tools have been extended to potential biosecurity threats for biodefense including botulinum toxin, Marburg and Ebola virus, smallpox virus, influenza viruses, coronaviruses, and agricultural plant pathogens such as *Brucella* (Amon et al. 2001; Madden and Wheelis 2003; Leffel and Reed 2004; Gopaul et al. 2014) as well as emerging and common foodborne and hospital pathogens (Hansen and Hendriksen 2001; Tauxe 2002; Bélanger et al. 2003; Toma et al. 2003; Ward and Bej 2006; Kagkli et al. 2012; Harrison and Hanson 2017; Hoseinpour et al. 2017; Li et al. 2017; Liu et al. 2018; Omer et al. 2018; Sánchez-Chica et al. 2020; Tastanova et al. 2021).

6.6 Microbiome Analysis

Microorganisms live on and in the human body. The 10-year human microbiome project (HMP) employed sequencing methods to characterize microbes that colonize the skin and live in the oral, nasal, urogenital, and gastrointestinal regions of the human body (Proctor 2016, 2019; The Integrative HMP (iHMP) Research Network Consortium 2019). Over 100 such species have been identified and characterized. Microbiome analysis spurred a new approach to body fluid, body region, and fingerprint analysis. Microbiome analysis of human body regions and body fluids have been used to profile the microbial composition at different body locations and differentiate lifestyle choice impacts on the microbiome (e.g., smoking, diet, living environment, pathobiology, physiology, host immune response, etc.) (Turnbaugh et al. 2007).

For example, identifying if a smoker or non-smoker contributed a fingerprint or body fluid to a crime scene can eliminate possible contributors. The genes CYP1A1, F2RL3, CYTL1, CYP1B1, and AHRR have been shown to be differentially methylated in smokers and nonsmokers in previous studies. Research has been shown that smoking can affect the methylation of DNA by suppressing gene expression or upregulating gene expression (Tsai et al. 2018). Upregulation of gene CYP1A1 can predict smoking status 90% of the time (Tekpli et al. 2011). The gene F2RL3 can differentiate smokers by hypomethylation (Zhang et al. 2014). DNA methylation at loci can be determined by pyrosequencing as described in Chap. 2. Caution must be used when evaluating DNA methylation, though, as smokers' DNA has been shown to revert to a methylation pattern of nonsmokers upon cessation of smoking (Tsaprouni et al. 2014). The microbiome has also been analyzed using samples taken from the front and back of cell phones (Lax et al. 2015).

The microbiome of soil and footwear soles has also been analyzed. The soil microbiome has been analyzed for forensic geolocation applications. Soil can be analyzed by morphological and chemical analysis to describe the color, density, pH, metal ion composition, chemical contaminants, and humic mass but these features may not be sufficient to individualize or classify the evidence or may require numerous assays including SEM,

TEM, GC-MS, XRD, and ICP-OES (Pirrie et al. 2017). Whole genome amplification followed by sequencing led to the differentiation of samples from two residential parks with soils of similar color (Khodakova et al. 2014). Analysis of the 16S and 18S rRNA gene hypervariable regions as well as whole genome sequencing using next generation sequencing have been used to fingerprint soil and water samples. In a study using the 18S rRNA gene, the soil microbiome has been shown to differentiate environments such as an urban park, grassland, fields, and forests (Lilje et al. 2013). Hjelmsø et al. (2014) used soil as a screening tool for bacterial communities using 16S rRNA sequencing. Recent research in this area has been focused on soil geolocation from Mid-Atlantic states in the United States using 16S rRNA (Malbrough and Elkins 2023). Analysis of the microbiome of footwear soles and cell phones using the 16S rRNA gene demonstrated that the microbiomes were different and that the microbiomes from the footwear of two people were different (Lax et al. 2015).

6.7 Conclusion

Non-human DNA and human DNA can both aid investigators in solving crimes. Non-human DNA can be used to solve cases involving bioterror agents or intentionally placed foodborne pathogens, animal perpetrators, and plant and fungal species or soil transferred from the crime scene to the evidence. Non-human DNA forensics continues to be an active area of research.

References

Alahi MEE, Mukhopadhyay SC (2017) Detection methodologies for pathogen and toxins: a review. Sensors 1885:17

Aly SM, Sabri DM (2015) Next generation sequencing (NGS): a golden tool in forensic toolkit. *Archiwum Medycyny Sądowej i Kryminologii/Archives of Forensic Medicine and Criminology* 65(4):260–271

Amendt J, Richards CS, Campobasso CP, Zehner R, Hall MJR (2011) Forensic entomology: applications and limitations. Forensic Sci Med Pathol 7:379–392

Amon SS, Schecter R, Inglesby TV, Henderson DA, Bartlett JG, Ascher MS, Eitzen E, Fine AD, Hauer J, Layton M, Lillibridge S, Osterholm MT, O'Toole T, Parker G, Perl TM, Russell PK, Swerdlow DL, Tonat K (2001) Botulinum toxin as a biological weapon: medical and public health management. JAMA 285(8):1059–1070

Antil S, Abraham JS, Sripoorna S, Maurya S, Dagar J, Makhija S, Bhagat P, Gupta R, Sood U, Lal R, Toteja R (2023) DNA barcoding, an effective tool for species identification: a review. Mol Biol Rep 50(1):761–775. https://doi.org/10.1007/s11033-022-08015-7

Barta CE, Bolander B, Bilby SR et al (2017) In situ dark adaptation enhances the efficiency of DNA extraction from mature pin oak (*Quercus palustris*) leaves, facilitating the identification of partial sequences of the 18S rRNA and isoprene synthase (IspS) genes. Plants (Basel) 6(4):52. https://doi.org/10.3390/plants6040052

Bartlett SE, Davidson WS (1992) FINS (forensically informative nucleotide sequencing)—a procedure for identifying the animal origin of biological specimens. BioTechniques 12(3):408–411

Bélanger SD, Boissinot M et al (2003) Rapid detection of *Clostridium difficile* in feces by real-time PCR. J Clin Microbiol 41:730–734

Bellis C, Ashton KJ, Freney L, Blair B, Griffiths LR (2003) A molecular genetic approach for forensic animal species identification. Forensic Sci Int 134(2–3):99–108

Bender AC, Faulkner JA, Tulimieri K, Boise TH, Elkins KM (2020) High resolution melt assays to detect and identify *Vibrio parahaemolyticus*, *Bacillus cereus*, *Escherichia coli*, and *Clostridioides difficile* bacteria. Microorganisms 8(4):561. https://doi.org/10.3390/microorganisms8040561

Bowser K, Smirnova D, Rivers DB, Elkins KM (2023) Spectroscopic analysis of fly artifacts. In: Srivastava A, Elkins KM (eds) Vibrational spectroscopy in forensic evidence analysis. Taylor & Francis/CRC Press. (submitted)

Bravi CM, Liron JP, Mirol PM, Ripoli MV, Peral-Garcia P, Giovambattista G (2004) A simple method for domestic animal identification in Argentina using PCR-RFLP analysis of cytochrome b gene. Legal Med 6(4):246–251

Butler J (2005) Forensic DNA typing, 2nd edn. Elsevier Academic Press, San Diego

Caine L, Lima G, Pontes L, Abrantes D, Pereira M, Pinheiro MF (2006) Species identification by cytochrome b gene: casework samples. Int Congr Ser 1288:145–147

Campobasso CP, Linville JG, Wells JD, Introna F (2005) Forensic genetic analysis of insect gut contents. Am J Forensic Med Pathol 26(2):161–165. https://doi.org/10.1097/01.paf.0000163832.05939.59

Chamoun CA, Couric MS, Garrido RG, Moura-Neto RS, Oliveira-Costa J (2019) *In vitro* recovery and identification of Y-STR DNA from *Chrysomya albiceps*

(Diptera, Calliphoridae) larvae fed a decomposing mixture of human semen and ground beef. Genet Mol Res 18(1):gmr18189

Chan AP, Crabtree J, Zhao Q, Lorenzi H, Orvis J, Puiu D, Melake-Berhan A, Jones KM, Redman J, Chen G, Cahoon EB, Gedil M, Stanke M, Haas BJ, Wortman JR, Fraser-Liggett CM, Ravel J, Rabinowicz PD (2010) Draft genome sequence of the oilseed species Ricinus communis. Nat Biotechnol 28:951–959

Clarke M, Vandenberg N (2010) Dog attack: the application of canine DNA profiling in forensic casework. Forensic Sci Med Pathol 6(3):151–157

Coopman V, De Leeuw M, Cordonnier J, Jacobs W (2009) Suicidal death after injection of a Castor bean extract (*Ricinus communis* L). Forensic Sci Int 189:e13–e20

Cowan AF, Elkins KM (2018) Detection and identification of *Psilocybe cubensis* DNA using a real-time polymerase chain reaction high resolution melt (PCR-HRM) assay. J Forensic Sci 63:1500–1505

Cowan AF, Elkins KM (2020) Detection and identification of kratom (*Mitragyna speciosa*) and marijuana (*Cannabis sativa*) by a real-time PCR high resolution melt duplex assay. J Forensic Sci 65:52–60

Craft KJ, Owens JD, Ashley MV (2007) Application of plant DNA markers in forensic botany: genetic comparison of *Quercus* evidence leaves to crime scene trees using microsatellites. Forensic Sci Int 165(1): 64–70. https://doi.org/10.1016/j.forsciint.2006.03.002

Deedrick DW, Koch SL (2004) Microscopy of hair part II: a practical guide and manual for animal hairs. Forensic Sci Commun 6(3):1–32

Despott E, Cachia MJ (2004) A case of accidental ricin poisoning. Malta Med J 16(4):39–41

Druml B, Cichna-Markl M (2014) High resolution melting (HRM) analysis of DNA—its role and potential in food analysis. Food Chem 158:245–254. https://doi.org/10.1016/j.foodchem.2014.02.111

Durdle A et al (2013) The human DNA content in artifacts deposited by the blowfly *Lucilia cuprina* fed human blood, semen and saliva. Forensic Sci Int 233(1–3): 212–219. https://doi.org/10.1016/j.forsciint.2013.09.015

Dwivedi HP, Jaykus L (2011) Detection of pathogens in foods: the current state-of-the-art and future directions. Crit Rev Microbiol 37:40–63

Eichmann C, Berger B, Reinhold M, Lutz M, Parson W (2004) Canine-specific STR typing of saliva traces on dog bite wounds. Int J Legal Med 118(6):337–342

Elkins KM (2019) Introduction to forensic chemistry. Taylor and Francis/CRC Press, Boca Raton, FL

Elkins K, Bender A (2020) Foodborne pathogens. Encyclopedia. https://encyclopedia.pub/512. Accessed 4 Jul 2022

Elkins KM, Marlette K (2023) Spectroscopic analysis of counterfeit wood products and application of forensic botany to solve criminal cases. In: Srivastava A, Elkins KM (eds) Vibrational spectroscopy in forensic evidence analysis. Taylor & Francis/CRC Press. (in press)

Elkins KM, Perez ACU, Sweetin KC (2016) Rapid and inexpensive species differentiation using a multiplex real-time PCR high resolution melt (HRM) assay. Anal Biochem 500:15–17

Elkins KM, Perez ACU, Quinn AA (2017) Simultaneous identification of four "legal high" plant species in a multiplex PCR high resolution melt assay. J Forensic Sci 62:593–601

Felder E, Mossbrugger I, Lange M, Wölfel R (2012) Simultaneous detection of ricin and abrin DNA by real-time PCR (qPCR). Toxins 4:633–642

Figueroa A, Ganitano DA, Solano JH, Anabalon LI (2017) Botanical DNA evidence in a case of illegal drug trafficking: the use of high-resolution melting (HRM) with the internal transcribed spacer (ITS) approach in the identification of psychedelic fungus (*Psilocybe sp.*). In: Proceedings of the 69th Annual Meeting of the American Academy of Forensic Sciences, New Orleans, LA, 13–18 Feb 2017. p 435

Gopaul KK, Sells J, Lee R, Beckstrom-Sternberg SM, Foster JT, Whatmore AM (2014) Development and assessment of multiplex high resolution melting assay as a tool for rapid single-tube identification of five *Brucella* species. BMC Res Notes 7:1–12

Greenberg B (1991) Flies as forensic indicators. J Med Entomol 28(5):565–577

Hansen BM, Hendriksen NB (2001) Detection of enterotoxic *Bacillus cereus* and *Bacillus thuringiensis* strains by PCR analysis. Appl Environ Microbiol 67: 185–189

Harrison LB, Hanson ND (2017) High-resolution melting analysis for rapid detection of sequence type 131 *Escherichia coli*. Antimicrob Agents Chemother 61:e00265-17

Henry P, Khatodia S, Kapoor K et al (2020) A single nucleotide polymorphism assay sheds light on the extent and distribution of genetic diversity, population structure and functional basis of key traits in cultivated north American cannabis. J Cannabis Res 2:26. https://doi.org/10.1186/s42238-020-00036-y

Hipp AL, Whittemore AT, Garner M, Hahn M, Fitzek E, Guichoux E, Cavender-Bares J, Gugger PF, Manos PS, Pearse IS, Cannon CH (2019) Genomic identity of white oak species in an eastern North American Syngameon. Ann Mo Bot Gard 104(3):455–477

Hjelmsø MH, Hansen LH, Bælum J, Feld L, Holben WE, Jacobsen CS (2014) High-resolution melt analysis for rapid comparison of bacterial community compositions. Appl Environ Microbiol 80:3568–3575

Hoseinpour F, Foroughi A, Nomanpour B, Nasab RS (2017) Identification and differentiation of *Campylobacter* species by high-resolution melting curve analysis. Microb Pathog 108:109–113. https://doi.org/10.1016/j.micpath.2017.05.009

Irwin A (2019) Tree sleuths are using DNA tests and machine vision to crack timber crimes. Nature 568: 19–21. https://doi.org/10.1038/d41586-019-01035-7

Kagkli D, Folloni S et al (2012) Towards a pathogenic *Escherichia coli* detection platform using multiplex SYBR green real-time PCR methods and high resolution melting analysis. PLoS One 7:e39287

Kanthaswamy S, Premasuthan A, Ng J, Satkoski J, Goyal V (2012) Quantitative real-time PCR (qPCR) assay for human-dog-cat species identification and nuclear DNA quantification. Forensic Sci Int Genet 6:290–295

Khodakova AS, Smith RJ, Burgoyne L, Abarno D, Linacre A (2014) Random whole metagenomic sequencing for forensic discrimination of soils. PLoS One 9(8):e104996. https://doi.org/10.1371/journal.pone.0104996.6

Kiesel B, Elkins KM (2019) Development of a PCR high resolution melt assay for *Artemisia absinthium* (wormwood) and a triplex assay with two additional "legal high" species *Datura stramonium* (jimson weed) and *Merremia tuberosa* (Hawaiian woodrose). J Forensic Sci 64:1817–1822

Kikkawa HS, Tahara M, Sugita R (2015) Forensic DNA analysis of wheat flour as commonly used in white powder cases. J Forensic Sci 60(5):1316–1321. https://doi.org/10.1111/1556-4029.12789

Kitpipit T, Chotigeat W, Linacre A, Thanakiatkrai P (2014) Forensic animal DNA analysis using economical two-step direct PCR. Forensic Sci Med Pathol 10(1):29–38

Knorr-Evans M, Parker-Forney M (2019) Buyer beware: one study finds 62% of wood products mislabeled. World Resources Institute, Washington, DC. https://www.wri.org/insights/buyer-beware-one-study-finds-62-wood-products-mislabeled. Accessed 11 Nov 2022

Kolavic SA, Kimura A, Simons SL, Slutsker L, Barth S, Haley CE (1997) An outbreak of *Shigella dysenteriae* type 2 among laboratory workers due to intentional food contamination. JAMA 278(5):396–398

Law JW, Mutalib NA et al (2015) Rapid methods for detection of foodborne bacterial pathogens: principles, applications, advantages and limitations. Front Microbiol 5:770

Lax S, Hampton-Marcell JT, Gibbons SM, Colares GB, Smith D, Eisen JA, Gilbert JA (2015) Forensic analysis of the microbiome of phones and shoes. Microbiome 3:21. https://doi.org/10.1186/s40168-015-0082-9

Leffel EK, Reed DS (2004) Marburg and Ebola viruses as aerosol threats. Biosecurity and bioterrorism: biodefense strategy. Pract Sci 2(3):186–191

Lemes MR, Brondani RPV, Grattapaglia D (2002) Multiplexed systems of microsatellite markers for genetic analysis of mahogany, *Swietenia* macrophylla King (meliaceae), a threatened neotropical timber species. J Heredity 93(4):287–290. https://doi.org/10.1093/jhered/93.4.287

Li X, Cai JF, Guo YD, Xiong F, Zhang L, Feng H, Meng FM, Fu Y, Li JB, Chen YQ (2011) Mitochondrial DNA and STR analyses for human DNA from maggots' crop contents: a forensic entomology case from central-southern China. Trop Biomed 28(2):333–338

Li B, Liu H, Wang W (2017) Multiplex real-time PCR assay for detection of Escherichia coli O157:H7 and screening for non-O157 Shiga toxin-producing E. coli. BMC Microbiol 17:215. https://doi.org/10.1186/s12866-017-1123-2

Lilje L, Lillsaar T, Rätsep R, Simm J, Aaspõllu A (2013) Soil sample metagenome NGS data management for forensic investigation. Forensic Sci Int Genet Suppl Ser 4:e35–e36. https://doi.org/10.1016/j.fsigss.2013.10.017

Linville JG, Hayes J, Wells JD (2004) Mitochondrial DNA and STR analyses of maggot crop contents: effect of specimen preservation technique. J Forensic Sci 49(2):341–344

Liu Y, Singh P, Mustapha A (2018) High-resolution melt curve PCR assay for specific detection of E. coli O157:H7 in beef. Food Control 86:275–282. https://doi.org/10.1016/j.foodcont.2017.11.025

Lutz L, Zehner R, Verhoff MA, Bratzke H, Amendt J (2021) It is all about the insects: a retrospective on 20 years of forensic entomology highlights the importance of insects in legal investigations. Int J Legal Med 135:2637–2651

Madden LV, Wheelis M (2003) The threat of plant pathogens as weapons against U.S. Crops Annu Rev Phytopathol 41:155–176

Malbrough J, Elkins KM (2023) NGS profiling of water and soil microbial DNA in forensic science. In: Dash HR, Elkins KM, Al-Snan NR (eds) NGS technology in DNA analysis. Elsevier. (in press)

Martinson EC, Peyton J, Kelkar YD, Jennings EC, Benoit JB, Werren JH, Denlinger DL (2019) Genome and ontogenetic-based transcriptomic analyses of the flesh Fly, *Sarcophaga bullata*. Genet Rep 9:1313

Meiklejohn KA, Jackson ML, Stern LA, Robertson JM (2018) A protocol for obtaining DNA barcodes from plant and insect fragments isolated from forensic-type soils. Int J Legal Med 132(6):1515–1526. https://doi.org/10.1007/s00414-018-1772-1

Meselson M, Guillemin J, Hugh-Jones M, Langmuir A, Popova I, Shelokov A, Yampolskaya O (1994) The Sverdlovsk anthrax outbreak of 1979. Science 266:1202–1208

Mohamed R, Tan HY, Siah CH (2012) A real-time PCR method for the detection of trnL-trnF sequence in agarwood and products from *Aquilaria* (Thymelaeaceae). Conserv Genet Res 4(3):803–806. https://doi.org/10.1007/s12686-012-9648-z

Müller K, Brugger C, Klein R, Miltner E, Reuther F, Wiegand P (2008) STR typing of hairs from domestic cats. Forensic Sci Int Genet Suppl Ser 1(1):607–609

Murch RS (2005) "Amerithrax": the investigation of bioterrorism using Bacillus anthracis spores in mailed letters. In: Encyclopedia of bioterrorism defense. Wiley, New York

Nakaki S-I, Hino D, Miyoshi M, Nakayama H, Moriyoshi H, Morikawa T et al (2007) Study of animal species (human, dog and cat) identification using a multiplex single-base primer extension reaction in the cytochrome b gene. Forensic Sci Int 173(2–3):97–102

Nithaniyal S, Newmaster SG, Ragupathy S, Krishnamoorthy D, Vassou SL, Parani M (2014) DNA barcode authentication of wood samples of threatened and commercial timber trees within the tropical dry evergreen forest of India. PLoS One 9(9): e107669. https://doi.org/10.1371/journal.pone.0107669

Njau DG, Muge EK, Kinyanjui PW, Omwandho COA, Mukwana S (2016) STR analysis of human DNA from maggots fed on decomposing bodies: assessment of the time period for successful analysis. Egypt J Forensic Sci 6:261–269. https://doi.org/10.1016/j.ejfs.2015.04.0024

Nugent KG, Saville BJ (2004) Forensic analysis of hallucinogenic fungi: a DNA-based approach. Forensic Sci Int 140:147–157

Omer MK, Álvarez-Ordoñez A, Prieto M, Skjerve E, Asehun T, Alvseike OA (2018) A systematic review of bacterial foodborne outbreaks related to red meat and meat products. Foodborne Pathog Dis 15:598–611

Pakull B, Schindler L, Mader M, Kersten B, Blanc-Jolivet-C, Paulini M, Lemes MR, Ward SE, Navarro CM, Cavers S, Sebbenn AM, di Dio O, Guichoux E, Degen B (2020) Development of nuclear SNP markers for Mahogany (*Swietenia* spp.). Conserv Genet Resour 12:585–587. https://doi.org/10.1007/s12686-020-01162-8

Parson W, Pegoraro K, Niederstatter H, Foger M, Steinlechner M (2000) Species identification by means of the cytochrome b gene. Int J Legal Med 114(1):23–28

Pavlovic M, Luze A, Konrad R, Berger A, Sing A, Busch U, Huber I (2011) Development of a duplex real-time PCR for differentiation between E. coli and Shigella spp. J Appl Microbiol 110:1245–1251

Pirrie D, Dawson L, Graham G (2017) Predictive geolocation: forensic soil analysis for provenance determination. Episodes 40(2):141–147. https://doi.org/10.18814/epiiugs/2017/v40i2/017016.8

Priyanka B, Patil RK et al (2015) A review on detection methods used for foodborne pathogens. Indian J Med Res 144:327–338

Proctor LM (2016) The National Institutes of Health human microbiome project. Semin Fetal Neonatal Med 21:368–372

Proctor L (2019) Priorities for the next 10 years of human microbiome research. Nature 569:623–625

Rajapaksha P, Elbourne AJ et al (2018) A review of methods for the detection of pathogenic microorganisms. Analyst 144:396–411

Rasko DA, Worsham PL, Abshire TG, Stanley ST, Bannan JD, Wilson MR, Langham RJ, Decker RS, Jiang L, Read TD, Phillippy AM, Salzberg SL, Pop M, Van Ertg MN, Keneficg LJ, Keim PS, Fraser-Liggett CM, Ravel J (2011) Bacillus anthracis comparative genome analysis in support of the Amerithrax investigation. PNAS 108(12):5027–5032

Reese KR, Elkins KM (2022) Simultaneous detection of foodborne pathogens using a real-time PCR triplex high-resolution melt assay. Appl Microbiol 2(3): 453–459. https://doi.org/10.3390/applmicrobiol2030034

Rivers D, Geiman T (2017) Insect artifacts are more than just altered bloodstains. Insects 8(2):37. https://doi.org/10.3390/insects8020037

Ryan AL, O'Hern CP, Elkins KM (2020) Evaluation of two new methods for DNA extraction of "legal high" plant species. J Forensic Sci 65(5):1704–1708. https://doi.org/10.1111/1556-4029.14478

Sánchez-Chica J, Correa MM et al (2020) A novel method for direct detection of *Bacillus cereus* toxin genes in powdered dairy products. Int Dairy J 103:104625

Sato I, Nakaki S, Murata K, Takeshita H, Mukai T (2010) Forensic hair analysis to identify animal species on a case of pet animal abuse. Int J Legal Med 124(3): 249–256

Sundqvist A, Ellegren H, Vila C (2008) Wolf or dog? Genetic identification of predators from saliva collected around bite wounds on prey. Conserv Genet 9: 1275–1279

Switai TL, Winter KJ et al (2015) Diagnosis and management of foodborne illness. Am Fam Physician 92:358–365

Tamburro M, Ripabelli G (2017) High resolution melting as a rapid, reliable, accurate and cost-effective emerging tool for genotyping pathogenic bacteria and enhancing molecular epidemiological surveillance: a comprehensive review of the literature. Ann Ig 29: 293–316. https://doi.org/10.7416/ai.2017.2153

Tanaka S, Ito M (2020) DNA barcoding for identification of agarwood source species using trnL-trnF and matK DNA sequences. J Nat Med 74(1):42–50. https://doi.org/10.1007/s11418-019-01338-z

Tang K, Ren J, Cronn R, Erickson DL, Milligan BG, Parker-Forney M, Spouge JL, Sun F (2018) Alignment-free genome comparison enables accurate geographic sourcing of white oak DNA. BMC Genomics 19(1):896. https://doi.org/10.1186/s12864-018-5253-1

Tarditi CR, Grahn RA, Evans JJ, Kurushima JD, Lyons LA (2011) Mitochondrial DNA sequencing of cat hair: an informative forensic tool. J Forensic Sci 56(Suppl. 1):S36–S46. https://doi.org/10.1111/j.1556-4029.2010.01592.x

Tastanova A, Stoffel CI, Dzung A, Cheng PF, Bellini E, Johansen P et al (2021) A comparative study of real-time RT-PCR-based SARS-CoV-2 detection methods and its application to human-derived and surface swabbed material. J Mol Diagn 23:796–804. https://doi.org/10.1016/j.jmoldx.2021.04.009

Tauxe RV (2002) Emerging foodborne pathogens. Int J Food Microbiol 78:31–41. https://doi.org/10.1016/s0168-1605(02)00232-5

Tekpli X, Zienolddiny S, Skaug V, Stangeland L, Haugen A, Mollerup S (2011) DNA methylation of the CYP1A1 enhancer is associated with smoking-induced genetic alterations in human lung. Int J Cancer 131:227421. https://doi.org/10.1002/ijc.27421

The Integrative HMP (iHMP) Research Network Consortium (2019) The integrative human microbiome project. Nature 569:641–648

Tobe SS, Linacre A (2010) DNA typing in wildlife crime: recent developments in species identification. Forensic Sci Med Pathol 6(3):195–206

Toma C, Lu Y et al (2003) Multiplex PCR assay for identification of human diarrhaegenic *Escherichia coli*. J Clin Microbiol 41:2669–2671

Török TJ, Tauxe RV, Wise RP, Livengood JR, Sokolow R, Mauvais S, Birkness KA, Skeels MR, Horan JM, Foster LR (1997) A large community outbreak of salmonellosis caused by intentional contamination of restaurant salad bars. JAMA 278:389–395

Tsai PC, Glastonbury CA, Eliot MN, BollePalli S, Yet I, Castillo-Fernandez JE et al (2018) Smoking induces coordinated DNA methylation and gene expression changes in adipose tissue with consequences for metabolic health. Clin Epigenetics 10:805580. https://doi.org/10.1186/s13148-018-0558-0

Tsaprouni LG, Yang TP, Bell J, Dick KJ, Kanoni S, Nisbet J et al (2014) Cigarette smoking reduces DNA methylation levels at multiple genomic loci but the effect is partially reversible upon cessation. Epigenetics 9:969637. https://doi.org/10.4161/15592294.2014.969637

Turnbaugh P, Ley R, Hamady M et al (2007) The human microbiome project. Nature 449:804–810. https://doi.org/10.1038/nature06244

Ward LN, Bej AK (2006) Detection of *Vibrio parahaemolyticus* in shellfish by use of multiplexed real-time PCR with TaqMan fluorescent probes. Appl Environ Microbiol 72:2031–2042

Wells JD, Introna F, Di Vella G, Campobasso CP, Hayes J, Sperling FA (2001) Human and insect mitochondrial DNA analysis from maggots. J Forensic Sci 46(3):685–687. https://doi.org/10.1520/jfs15022j

Wiedenhoeft A (2016) Best practice guide for forensic timber identification. In: United Nations office on drugs and crime: international consortium on combating wildlife crime. Laboratory and Scientific Section, Global Programme for Combating Wildlife and Forest Crime, Vienna, pp 1–226. https://www.unodc.org/documents/Wildlife/Guide_Timber.pdf. Accessed 14 Oct 2022

Yoon C (1993) Forensic science. Botanical witness for the prosecution. Science 260(5110):894–895

Young JM, Weyrich LS, Cooper A (2014) Forensic soil DNA analysis using high-throughput sequencing: a comparison of four molecular markers. Forensic Sci Int Genet 13:176–184

Zaya DN, Ashley MV (2012) Plant genetics for forensic applications. Methods Mol Biol 862:35–52. https://doi.org/10.1007/978-1-61779-609-8_4

Zehner R, Amendt J, Krettek R (2004) STR typing of human DNA from fly larvae fed on decomposing bodies. J Forensic Sci 49(2):1–4. https://doi.org/10.1520/jfs2003248

Zhang Y, Yang R, Burwinkel B, Breitling LP, Brenner H (2014) F2RL3 methylation as a biomarker of current and lifetime smoking exposures. Environ Health Perspect 122:306937. https://doi.org/10.1289/ehp.1306937

Applications of NGS Technology in Forensic DNA Analysis

7

7.1 Introduction to Technology in Forensic DNA Typing

New technology is being applied to forensic DNA analysis to move samples from collection at the crime scene, to sampling, extraction of DNA, and genotyping of targeted loci. While the crime scene investigation, packaging, and sampling steps cannot be automated, there are several excellent commercial instruments for automating DNA extraction and DNA typing processes in the laboratory. These can be used to detect and identify DNA from human suspects as well as DNA from other species. Microbial and viral DNA may be encountered in bioterrorism crimes or used in lifestyle or soil geolocation analysis, plant DNA may be encountered in legal high drug cases and for geolocation applications, and animal DNA may be encountered in cases where pets and animals are the perpetrators or were at the crime scene. DNA extraction instruments include the Promega Maxprep, Maxwell 16, Maxwell FSC, and Maxwell 48 and Qiagen Biorobot EZ1, EZ1 Advanced, and EZ2 Connect. The Maxwell instruments utilize radio frequency identification (RFID) for detecting loaded reagents and a barcode reader can be used to track samples. Modern capillary electrophoresis (CE) instruments including the Applied Biosystems 3500 series and Promega Spectrum SE employ RFID readers and 6- and 8-dye detection, respectively. Next generation sequencing (NGS) instruments also employ RFID technology. Sample sheets for CE and NGS can be prepared and uploaded to the instrument or manually inputted on the day of the run. NGS library prep instruments include the Verogen PrepStation and ThermoFisher Ion Chef. NGS instruments include the Illumina iSeq 100, Illumina MiniSeq, Illumina MiSeq, Verogen MiSeq FGx, Illumina HiSeq, Ilumina NovaSeq, Thermo Fisher Ion Torrent, ThermoFisher Ion PGM, ThermoFisher Ion S5. Thermo Fisher Ion GeneStudio. DNA extraction robots, library preparation and pre-DNA typing handling robots, and DNA sequencing technology reduce the hands-on time, processing time and reduce manual handing errors. Several manufacturers sell commercial kits for forensic DNA analysis using these instruments as described in Chap. 1. The applications of NGS technology in forensic DNA analysis is the focus of this chapter.

7.2 DNA Extraction Technology

7.2.1 Maxprep, Maxwell 16, Maxwell FSC, and Maxwell 48 Instruments by Promega

The Maxwell 16, Maxwell FSC, and Maxwell 48 are compact robots that automate DNA extraction for bone, body fluid, cultured cells, tissue, and microbe samples. The samples are heated in lysis buffer with proteinase K in a shaking incubator for 2 h prior to loading on the instrument.

H. R. Dash et al., *Advancements in Forensic DNA Analysis*,
https://doi.org/10.1007/978-981-99-6195-5_7

Alternatively, the samples can be prepared by the maxprep liquid handler. The maxprep can add the lysis buffer and proteinase K. It can perform heating and shaking in the incubation step prior to extraction and transfer the samples to the cartridges. The instruments use cartridges that contain all of the reagents and elution solutions needed for DNA extraction. The user opens the seal on the cartridge to load the sample to be extracted and a plunger. The user loads the cartridge and elution tubes with the caps open to the Maxwell tray. One to 48 samples can be loaded for simultaneous extraction and only the cartridges needed for the extraction are used. The extraction uses paramagnetic particles. The extraction time by the instrument is 20 min. The extracted DNA can be used for qPCR and NGS applications. All of the instruments have an integrated UV for decontamination and a barcode reader for sample tracking except for the Maxwell 16 for which it is an option. All of the instruments integrate a tablet interface for guiding the user through setup, tracking the extractions including a run summary, and troubleshooting including detecting plugged tips. Integrating a Quantus fluorimeter for quantitation is an option with the Maxwell 48. The Maxprep can also be used for post-extraction tasks including adding dye and reagents for qPCR and normalization of NGS samples. The Maxwell FSC is shown in Fig. 7.1.

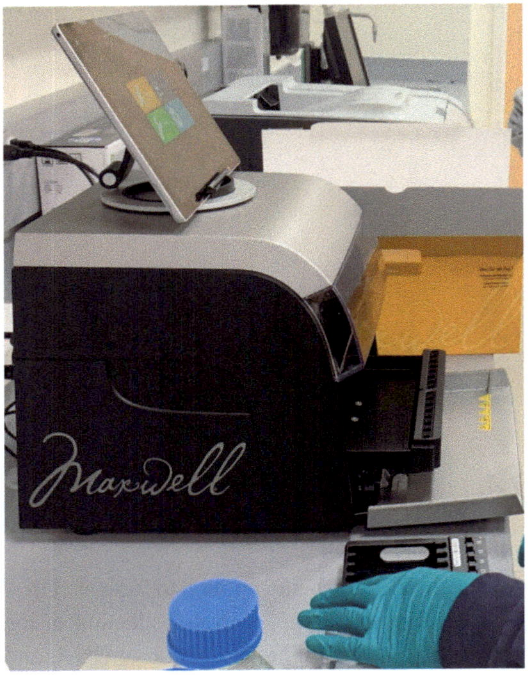

Fig. 7.1 Loading the Maxwell FSC Instrument

7.2.2 Biorobot EZ1, EZ1 Advanced, and EZ2 Connect Instruments by Qiagen

The Biorobot EZ1, EZ1 Advanced, and EZ2 Connect can extract DNA and RNA from samples including bone, body fluids, tissue, cells from culture, formalin-fixed paraffin-embedded (FFPE), and hair samples. Models can extract 6 or 24 samples simultaneously. Like the Maxwell systems, users load the number of cartridges corresponding to the number of samples to be extracted. Unlike the Maxwell, the cartridges do not need to be punctured by the user reducing the risk for loss of reagents or contamination from handling. Rather, an air pressure system punctures the foil seal in the first step in the instrument. The cartridges should be inverted a few times before using to resuspend the magnetic beads. A memory card inserted into the instrument contains the run parameters for the extraction. The user interface instructs the user to load the cartridges into the cartridge rack, tip holders, tips, elution tubes, close the hood, and press "start." Like the Maxwell series, the EZ1/EZ2 extraction uses magnetic particle technology, and a UV lamp decontaminates the robot following sample extraction. Following incubation with lysis buffer, proteinase K and DTT (for sperm samples), the extraction process takes 18 min. A barcode reader can be used with the instrument. The EZ2 Connect and Maxwell instruments both use touchscreen tablets to control the instrument operations and select sample lanes for extraction, record sample IDs and notes, and connect run information. The display shows the protocol progress and run completion and error information.

7.3 Capillary Electrophoresis Technology for DNA Typing

While the earliest capillary electrophoresis instruments such as the Applied Biosystems Prism 310 Genetic Analyzer were single capillary and were capable of 4-dye detection, modern instruments can have 16 capillaries in parallel and detect up to 8 dyes simultaneously. Among the newest instruments for capillary electrophoresis are the Applied Biosystems 3500 series and the Promega Spectrum SE with Hitachi High-Tech. These instruments can conduct Sanger sequencing and fragment analysis approaches to DNA genotyping for human identification and wildlife forensics. The 36-cm capillary generally is used for fragment analysis and 47 cm for sequencing applications.

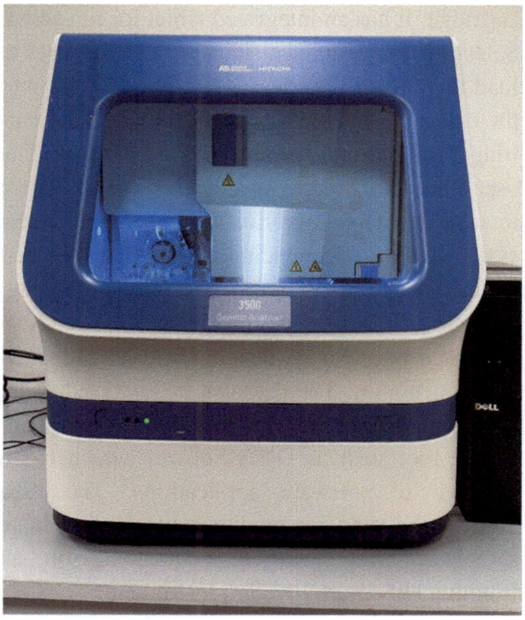

Fig. 7.2 Applied Biosystems 3500 Instrument

7.3.1 Applied Biosystems 3500

Applied Biosystems (AB) released its 3500 and 3500 XL Genetic Analyzer instruments in 2009. The Applied Biosystems 3500 instrument (Fig. 7.2) is an 8-capillary electrophoresis instrument capable of 6-dye detection that can perform DNA sequencing, fragment analysis, and human identification. The AB 3500 XL is a 24-capillary instrument. Users can use 96 or 384 well plates or individual tubes. There are also 3500 Dx and 3500 XL Dx models. The 3500 series has a smaller footprint than previous AB genetic analyzers and plugs into a standard power outlet. Two plates can be loaded into the instrument for processing. Its laser is a solid state 505 nm excitation source. Consumables, including polymer, are packaged in pouches with RFID tags. The anode and cathode buffers are also RFID tagged with lot numbers and expiration dates. The capillary array is a modular plug and play design. The 3500 series software wizard leads users through plate setup, troubleshooting, and analysis. Users can monitor runs and view data in the software and prioritize sample run order. Maintenance tasks can be scheduled as desired by users.

7.3.2 Promega Spectrum SE

The Spectrum SE instrument by Promega is an 8-capillary instrument capable of 8-dye detection released in 2021. It has a touchscreen tablet screen and integrated UV disinfection. Consumables are sold in single-use packs that are easy to load to the instrument and are tracked using RFID tags. These include a polymer cartridge, sample cartridge, and consumables. An attached barcode reader can be used to track samples. The run parameter and data analysis software can be accessed remotely. The Spectrum is also available in a compact model 6-dye benchtop instrument (Spectrum Compact SE) with a 32-sample capacity that was unveiled in 2017 at the American Academy of Forensic Sciences (AAFS) National Meeting. The top door opens to reveal where the consumables are loaded.

7.4 Liquid Handling Technology for NGS Library Prep

The liquid handling technology for automating library preparation steps for NGS include the PrepStation by Verogen and the Ion Chef by Life Technologies.

7.4.1 PrepStation by Verogen

The Verogen PrepStation liquid handler semi-automates library prep steps performed prior to NGS. The software can communicate with the MiSeq FGx sequencer and Universal Analysis System (UAS) post-sequencing tool. It can be used with the Verogen ForenSeq MainstAY, MainstAY SE, Imagen, and DNA Signature Prep kit for 8, 16, 24, 36, 40, or 48 reactions and the ForenSeq Kintelligence kit for 3, 6, or 12 reactions.

7.4.2 Ion Chef by Life Technologies

The Life Technologies Ion Chef system from ThermoFisher fully automates the library prep steps for the Ion AmpliSeq Panels performed prior to NGS complete to sequence-ready chip for the Ion Proton and Ion PGM systems. It can also be used to prepare microbiome, transcriptome, and exome libraries. To begin, the reagent cartridges, consumables including tubes and tips, plate with samples, and Ion semiconductor sequencing chips are loaded. Within it is a barcode scanner and RFID chips. The Ion Chef has within it a thermocycler to prepare templated ion sphere particles for sequencing. To a run, the user selects "run" and "PGM" or "proton". The chip is ready for sequencing following the Ion Chef run.

7.5 NGS Technology for DNA Typing

The next generation sequencing instruments include the Illumina iSeq, Illumina MiSeq, Verogen MiSeq FGx, Illumina NextSeq, Illumina HiSeq, Illumina NovaSeq, Thermo Fisher Ion Torrent, Thermo Fisher Ion PGM, and Thermo Fisher Ion S5. Several manufacturers including Promega, Thermo Fisher, Qiagen, and Verogen sell commercial kits for forensic DNA analysis applications with these instruments. The next generation sequencing instruments have few moving parts and plug and play reagents including sequencing chips or flow cells and reagent cartridges.

7.5.1 Illumina iSeq

The Illumina sequencers use a sequencing by synthesis (SBS) chemistry and cluster generation approach for paired index 151 bp length (2×151) runs. The Illumina iSeq 100 is one of the low-cost sequencers and has a small footprint that is even smaller than many thermocyclers. It has an integrated RFID to track consumables. The samples processed through library prep are pooled and sequenced on a flow cell. All of the sequencing reagents and wash steps are contained in a cartridge that is inserted into the instrument. The pooled libraries are loaded to a well in the cartridge. It has an integrated tablet for run setup. Selecting the "sequence" icon prompts the user to load the pooled libraries, insert the flow cell into the cartridge, opens the chamber to load the cartridge, and prompts the user to close the door. The user loads their sample sheet and can select where to save the data. Libraries are denatured by the system prior to sequencing eliminating the need for the user to perform this step prior to loading. The iSeq 100 can perform four million reads in 17 h. It is used for sequencing viruses and small whole genomes. The data can be analyzed using Illumina's Basespace applications or an open-source tool such as DNA Subway which has a package of software applications. The iSeq 100 can be used for quality control to evaluate sample pooling prior to sequencing on an Illumina HiSeq or NovaSeq instrument.

7.5.2 Illumina MiniSeq

The Illumina MiniSeq shares many similarities to the iSeq 100 with the compact design, 2 × 151 sequencing, integrated tablet, cartridge system and push button approach, and post-run analysis tools. It can be used to sequence samples prepared with the Nextera Rapid Capture, Nextera Library Preparation, TruSeq Custom Amplicon, TruSeq Targeted RNA Expression, and TruSeq Small RNA kits for resequencing, tumor profiling, RNA expression, and germline testing. The pooled samples are loaded to the reagent cartridge and the sequencing is begun by selecting "sequence" on the touchscreen, selecting the kit used, and loading the flow cell to its compartment. A waste container is placed in a waste and wash chamber. The MiniSeq can perform up to 50 million reads in approximately 24 h.

7.5.3 Illumina MiSeq

The Illumina MiSeq can be used for small genome sequencing such as for viruses and bacteria and targeted human, plant, and animal genome sequencing. It is a benchtop sequencer that performs 2 × 151 bp sequencing and uses 8 bp barcodes for demultiplexing. For each sequencing run the user loads the flow cell, reagent cartridge with pooled and denatured library samples, incorporation buffer, and waste container. The number of samples that can be sequenced simultaneously varies depending upon the primer set used, number of loci probed, and desired read depth. For human identification and using the Verogen kits, the upgraded MiSeq FGx model is needed and the instrument is run in Forensic Genomics mode. For all other applications including microbiome analysis, the MiSeq or MiSeq FGx research use only (RUO) mode is used. Up to 400 samples can be sequenced simultaneously for 16S rRNA gene targeted microbiome projects. The instrument is washed with dilute Tween detergent following each sequencing run and once a week unless the instrument is powered down.

7.5.4 Verogen MiSeq FGx

The Verogen MiSeq FGx (Fig. 7.3) is built on the MiSeq platform and performs 2 × 151 sequencing (Caratti et al. 2015). The Illumina MiSeq is inexpensive to run compared to other platforms, is easy to use, robust, and has a low error rate. The flow cells and cartridges for Verogen human identification applications can only be purchased from Verogen (now Qiagen). As on the Illumina MiSeq, the prepared sequencing libraries are loaded to the reagent cartridge and the reagent cartridge, flow cell, and waste container are inserted into the instrument. It can be used to perform targeted sequencing for human identification applications including targeted and whole mitochondrial genome sequencing, targeted STR and SNP sequencing, and targeted larger scale SNP loci sequencing. It can be used with the Verogen mitochondrial whole genome sequencing and control regions kits and the ForenSeq Signature Prep (Jäger et al. 2017; Guo et al. 2017), MainstAY (Stephens et al. 2023), Imagen (Elkins et al. 2023), Kintelligence (Snedecor et al. 2022) kits for human nuclear DNA typing. The number of samples that can be processed in a run varies; up to 96 samples can be sequenced simultaneously using the ForenSeq kit Primer Set A (Jäger et al. 2017) or the MainstAY kit Primer Set C (Beta test kit provided to the author) while only three samples can be sequenced simultaneously using the 10,230 loci Kintelligence kit on a standard flow cell (Verogen Demo data provided to the author). It can also be used to perform microbial, viral, and wildlife targeted ITS, COI, and 16S rRNA gene DNA sequencing using the Research Use Only (RUO) mode.

7.5.5 Thermo Fisher Ion S5

The Ion S5 can sequence microbial genomes, transcriptomes, gene panels, and exomes using different chips. It is also available in a S5XL model. The waste container is located in the left door of the instrument and the sequencing buffer

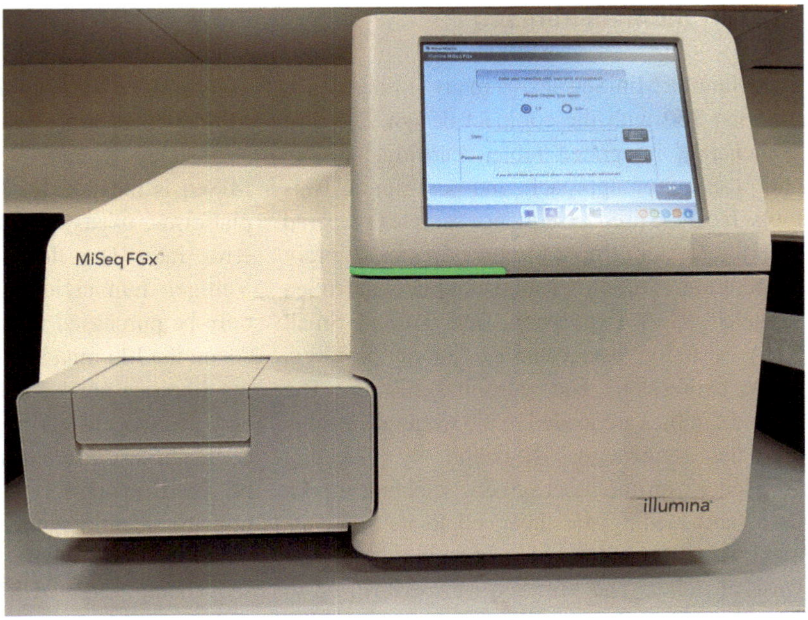

Fig. 7.3 Verogen MiSeq FGx Instrument

snaps into place in front of it. The cleaning solution is loaded into the door and the reagent cartridge with all of the reagents needed for sequencing is loaded into its drawer. All of the consumables are tracked using RFID tags. When initialization is complete, the sequencing chip is loaded into its compartment. It is compatible with the Ion 520 or 530 chips and can run up to 32 samples in a run. A sequencing run is completed in approximately 2.5 h. The ThermoFisher GlobalFiler™ NGS STR Panel kit and Precision ID GlobalFiler™ NGS STR Panel v2 have been validated on the S5 (Tao et al. 2019; Faccinetto et al. 2019).

7.5.6 ThermoFisher Ion Torrent PGM

The ThermoFisher Ion sequencers employ semiconductor chips. The chip contains millions of pixels with millions of wells covering the pixels. The digital information is converted to base calls. The Ion Torrent PGM can sequence bacterial genomes and conduct targeted gene sequencing and forensic NGS applications. Released in 2013, It is easy to use, sensitive, and generates data fast and can be used with the Ion HID-Ion AmpliSeq™ Identity targeted NGS loci panel of up to thousands of loci (Guo et al. 2016). Prior to sequencing on the Ion systems, the DNA is cut into millions of fragments. The fragments are attached to their own beads and copies until they cover the bead. Millions of beads are flowed across the chip and deposit into a well. The chip is flooded with each of the four DNA nucleotides. When a nucleotide is incorporated, a pH change is generated by the release of a hydrogen ion. A sensor records the change in pH and converts it to voltage. Every 15 s, a new nucleotide is introduced over the chip. When two identical bases are incorporated, the detected voltage is doubled. It can be used with 314 (50–100 Mb), 316 (up to 1 GB of sequence), and 318 (up to 2 GB of sequence) chips. Data can be viewed with the Converge software which calls sequence variants. The complete mitochondrial genome sequencing has also been tested on the Ion Torrent PGM™ (Zhou et al. 2016) and the Ion Torrent has be used with the ThermoFisher Precision ID Identity kit (Meiklejohn and Robertson 2017).

7.5.7 ThermoFisher Ion GeneStudio S5

ThermoFisher Ion GeneStudio S5 has a sequencing reagent cartridge, chip, sequencing buffer, and waste container bottle that are inserted by the user as in the MiSeq. It also uses RFID tags. It can be used to analyze microbes, exomes, and transcriptomes in 2.5 h. It can be used with five different semiconductor chip types including the 510, 520, 530, 540, and 550 and is also compatible with the ThermoFisher GlobalFiler™ NGS STR Panel kit.

7.5.8 Illumina NextSeq

The Illumina NextSeq series includes 500, 1000, and 2000 models. Sequencing runs can be remotely or locally set up. The reagent cartridge with prepared libraries, resuspension buffer, and flow cell(s) are loaded to the instrument. Sequencing is begun by selecting "sequence" on the attached touchscreen tablet. It can be used to sequence one whole human genome, 9 exomes, 10 transcriptomes, or 40 mRNA gene expression profiles for up to 400 million reads. It can be used to detect circulating tumor cells in the blood or to sequence thousands of organisms in parallel to identify low-abundance or unculturable microbes or evaluate microbial diversity using shotgun metagenomic sequencing. It can be used to sequence the whole genome of a bacterium, virus, or other microbe or conduct deep 16S rRNA gene sequencing.

7.5.9 Illumina HiSeq

The Illumina HiSeq performs 2 × 75, 2 × 125, 2 × 151, and 2 × 251 bp sequencing. There are HiSeq 3000, 4000 and X 10 series instruments. Using the Illumina HiSeq X 10, a whole human genome can be sequenced at 30× coverage for under $1000 USD in consumables. It consists of 10 HiSeq X instruments. It yields up to six billion bases per run and up to 1800 Gb in data in less than 3 days. The HiSeq 3000 can sequence six human genomes per run while the HiSeq 4000 can sequence 12 human genomes per run in 3.5 days. The instrument can run two flow cells simultaneously. It uses patterned flow cell technology for increased cluster density and simplified imaging.

7.5.10 Illumina NovaSeq

The Illumina NovaSeq including the X Series and 6000Dx are large like a mainframe computer but are extremely powerful in their sequencing capabilities. They are intended for high throughput sequencing facilities and has an integrated tablet and computer server. The NovaSeq 6000Dx can sequence human whole genomes, whole exomes, and whole transcriptomes in 2 × 251 sequencing with 10 bp indices. The NovaSeq X Series can also sequence human whole genomes, whole exomes, and whole transcriptomes in 2 × 151 sequencing with 10 bp indices. The NovaSeq X Plus can sequence 128 human genomes at 30× coverage or 96 human genomes at 40× coverage using the dual flow cells in one run. The user inserts the prepared DNA library tube into the reagent cartridge and loads the SBS cartridge, cluster cartridge, buffer cartridge, and flow cell into the sequencer drawers within the double door sequencer cabinet. The flow cell is larger (approximately the length of a human hand) to sequence more libraries simultaneously. The consumables are all tracked using RFID tags. The sequencing run is begun by selecting "sequence" on the touch screen or by access remotely through a web browser. The user creates the new run in the Illumina sequencing managed by selecting "create run" and entering the run setting and sample data manually or by importing from a sample sheet. One or two flow cells can be selected for the run. Sequencing on the NovaSeq 6000Dx can take up to 44 h and output up to 6000 Gb of data including 40 billion bases and sequencing on the NovaSeq X Series can take up to 48 h, generate up to 16,000 Gb of data, and up to 104 billion

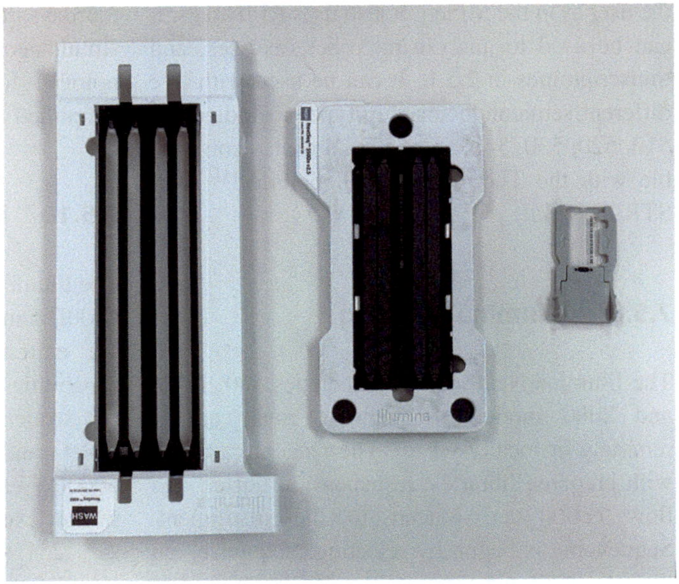

Fig. 7.4 Illumina Flow Cells for the (left to right) NovaSeq, NextSeq, and MiSeq

Table 7.1 Capillary electrophoresis genetic analyzer instruments for forensic applications

Instrument	Number of dyes	Number of capillaries	Company
310	4	1	Applied Biosystems
3100	5	16	Applied Biosystems
3100-*Avant*	5	4	Applied Biosystems
3130	5	4	Applied Biosystems
3130xl	5	16	Applied Biosystems
3500	6	8	Applied Biosystems
3500xl	6	24	Applied Biosystems
3730	5	48	Applied Biosystems
3730xl	5	96	Applied Biosystems
SeqStudio	6	4	Applied Biosystems
SeqStudio 8 Flex	6	8	Applied Biosystems
SeqStudio™ 24 Flex	6	24	Applied Biosystems
Spectrum Compact SE	6	4	Promega
Spectrum SE	8	8	Promega

paired-end reads using for NovaSeq X Plus system. A post-run wash is conducted by selecting "wash" on the touch screen. The data can be analyzed from the run file and using Illumina Basespace including the DRAGEN Germline, Enrichment, Somatic, and RNA tools. A comparison of Illumina flow cells is shown in Fig. 7.4.

7.5.11 Summary of CE and NGS DNA Typing Instruments

Table 7.1 summarizes the CE and NGS instruments for DNA typing fragment analysis and sequencing applications. In the past 25 years, capacity has increased from 1 to 24 capillaries and detection capability from 4 to

Table 7.2 Next generation sequencing instruments for forensic applications

Instrument	Number of samples	Company
MiSeq FGx	3–96	Verogen
Ion S5	32 each chip (up to 2 per run)	ThermoFisher
Ion PGM	1 or more per chip depending on how many target loci	ThermoFisher
HID Ion GeneStudio S5	32,96	ThermoFisher

8 fluorescent dyes. The MiSeq standard flow cell and Ion Torrent chips can sequence approximately three million bases. Table 7.2 summarizes NGS instruments for forensic applications.

7.6 Software Tools for CE and NGS Data Analysis

Following CE and NGS data collection, there are several software applications and packages that can be used to process the data. The GeneMapper™ ID-X software can be used to view sizing data and allele calls for CE fragment analysis. Data collected using commercial short tandem repeat (STR) kits and user-designed assays can be analyzed. Users can create new projects and set thresholds and evaluate sizing and ladder data. Off-ladder alleles can be assigned manually based on inspection and comparison to the allelic ladder.

NGS produces more data and requires steps for raw data QA and analysis than CE. NGS data undergoes several preprocessing steps following collection. After preprocessing, taxonomic profiling and predictive metagenomics profiling can be conducted. Preprocessing includes genome assembly, quality control, and chimera detection. For 16S rRNA gene sequence data analysis of microbiomes in body fluids and human skin samples or soil DNA extracts for geolocation, several applications are applied. Illumina Basespace has an array of processing and interpretation tools. The PANDAseq algorithm can be used to assemble the sequencing data after barcode demultiplexing by comparison, alignment, and scoring to a reference genome sequence. Mismatches and short alignment length sequences have low scores and are removed. Adaptor and PCR primer sequences and low quality sequences are also removed based on the Phred quality score. For example, a Phred quality score of 10 refers to a 90% accuracy for the base call (1 error expected for every 10 bases, on average) while a quality score of 60 indicates a 99.9999% accuracy (or 1 in 1,000,000 probability of incorrect base call). Removal of sequences with low quality scores improves the accuracy of allele calls. Applications for checking for trimming and checking for quality reads include FastQC, Qualitative Insights into Microbial Ecology (QIIME), SILVAangs, and the Brazilian Microbiome Project Operating System (BMPOS). Chimeras formed during PCR amplification can arise following incomplete extension of two loci which become primers and lead to a chimera product. Following taxonomic assignment with a tool such as USEARCH, chimeras can be detected by UCHIME and removed. Taxonomic profiling includes sequence alignment, assessment of microbial composition, and diversity analysis. Sequences can be clustered to phylotypes or operational taxonomic units (OTUs). The databases for 16S rRNA gene annotation include the human microbiome project (HMP), ribosomal database project (RDP), greengenes (https://greengenes.lbl.gov/), and SILVA (https://www.arb-silva.de/). Following taxonomic profiling, microbial diversity and composition can be visualized using pie charts and bar charts. Alpha (α) diversity refers to species richness within a sample or system and beta (β) diversity refers to the diversity in different environments. The OTUs can also be used in predictive metabolic profiling to understand the role of the microbiome on host metabolism and disease. Tools include PICRUSt, Tax4Fun, MicrobiomeAnalyst, and Piphillin. The bioinformatics pipelines commonly used for 16S rRNA gene analysis include MetaGenome Rapid Annotation using Subsystem Technology (MG-RAST), QIIME, and Mothur. The three pipelines have been shown to produce comparable results.

QIIME has been implemented in the cloud-based DNA Subway (https://dnasubway.cyverse.org/) NGS bioinformatics platform. The "Subway" has five lines: red, yellow, blue, green, and purple. The red line has tools to predict genes, search databases, and annotate plant and animal genomes up to 150 kb. The yellow line can be used to perform a TARGeT search to identify homologs using a tree and alignment viewer. The blue line will assemble DNA sequences and output phylogenetic trees. The green line can be used to examine RNA-Seq data to evaluate differential gene expression and transcriptomes. The purple line can be used to evaluate environmental DNA samples using QIIME and evaluate alpha and beta diversity.

For human DNA sequence analysis, similar research pipelines and approaches can be used especially when focusing on a gene of interest for disease genotyping and using RUO mode. Open-source applications for human DNA STR and SNP analysis in NGS data include STRSeq, STRaitRazr, FDSTools, and PIRANHA. STRSeq is a database of sequence variant diversity at human identification STR loci (Gettings et al. 2017). FDSTools can be used to identify the number of sequenced loci in FastQ and FastA files provided and count the number of times each unique sequence is detected in each sample, recognize and quantify stutter artifacts, and predict stutter (Hoogenboom et al. 2017). STRaitRazr is a tool in R that can be used to analyze human NGS STR, SNP, insertion/deletions, and mitochondrial DNA data and detect severe heterozygote peak imbalance (King et al. 2021). The French National Police developed the Programme d'Interprétation Résultats d'Analyses NGS Hautement Amélioré (PIRANHA) in R that reads the UAS summary report text file to further interpret the ForenSeq data.

Alternatively, for analyzing STR and SNP loci throughout the genome using commercial forensic NGS DNA Typing kits, several commercial products have been developed for the data analysis that reduce the need to toggle between spreadsheets, apps and tools. These include the Verogen Universal Analysis System (UAS), QIAGEN CLC Genomics Workbench, and ThermoFisher Converge Software NGS Analysis Module. The CLC Genomics Workbench can analzye RNA-seq and small RNA expression data, assemble for large eukaryotic genomes, perform de novo prediction of eukaryotic genes, perform de novo assembly of genomes and transcriptomes, perform methylation bisulfite and NGS variant calling and can be used with Illumina and ThermoFisher Ion Torrent data. The UAS calls STR and single nucleotide polymorphism (SNP) genotypes (Fig. 7.5) and uses the SNP sequence data to predict the biogeographical ancestry and phenotype characteristics including skin tone, eye color, and hair color (Fig. 7.6). It analyzes and displays the human sequencing control and positive and negative control results and outputs and graphs the overall read count for the controls and samples. Each STR and SNP locus can be probed for number of reads for each allele and imbalance and sequence. Flanking region sequence data files can be downloaded. The UAS can be used to compare samples side-by-side and compute random match probability (RMP) statistics. Kintelligence files (Fig. 7.7) can be viewed and downloaded for upload to GEDmatch for forensic investigative genetic genealogy (FIGG) applications. Like the UAS, Converge displays calls STR alleles and SNP variants and displays the number of reads and sequences for each allele. Flanking region reports can be generated. It can also be used for case

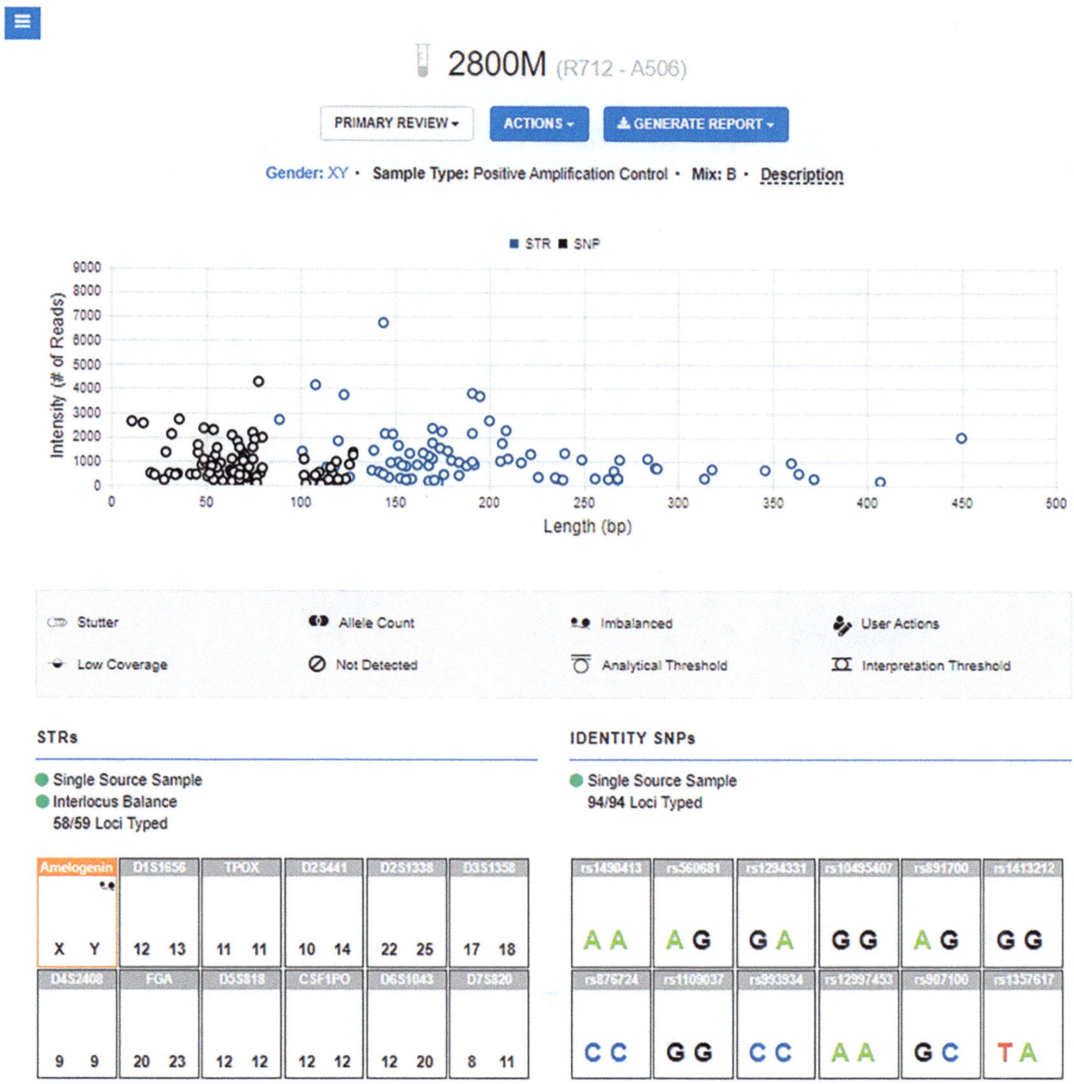

Fig. 7.5 Verogen UAS STR calls

management, kinship and paternity analysis, ancestry predictions, and to compare CE and NGS data and create a composite profile. Users can create pedigree trees and calculate likelihood ratios (LR), probability of exclusion (PE) and random man not excluded (RMNE) statistics. Photos and documents can be uploaded into reports for case management.

7.7 Conclusion

NGS instruments and technology have infiltrated forensic DNA analysis in the past 10 years. Targeted and whole microbial, viral, wildlife, and human genomes can be sequenced using these technologies. Advancements in statistical and probabilistic genotyping analyses, DNA databases, and real-time case work applications of these tools are the focus of the chapters that follow.

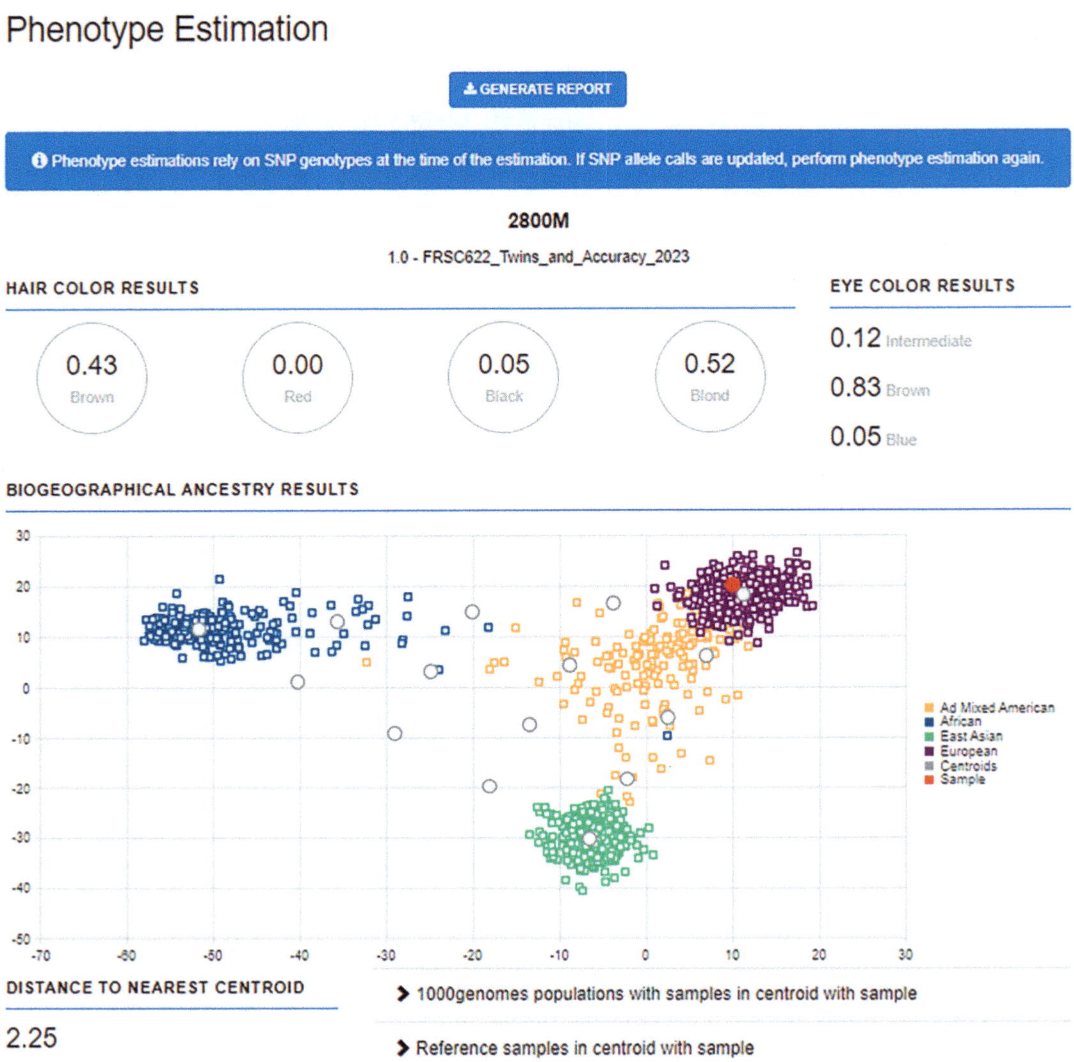

Fig. 7.6 Verogen UAS Phenotype and biogeographical ancestry prediction

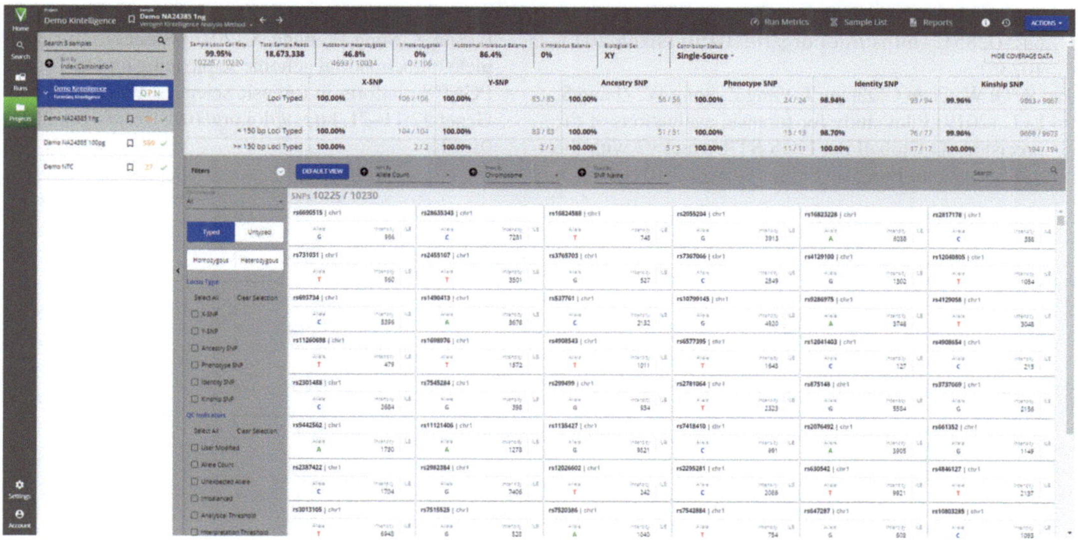

Fig. 7.7 Verogen Kintelligence results

References

Caratti S, Turrina S, Ferrian M et al (August 2015) MiSeq FGx sequencing system: a new platform for forensic genetics. Forensic Sci Int Genet Suppl Ser 5:e98–e100. https://doi.org/10.1016/j.fsigss.2015.09.040

Elkins KM, Garloff AT, Zeller CB (2023) Additional predictions for forensic DNA phenotyping of externally visible characteristics using the ForenSeq and Imagen kits. J Forensic Sci 68(2):608–613. https://doi.org/10.1111/1556-4029.15215

Faccinetto C, Serventi P, Staiti N, Gentile F, Marino A (2019) Internal validation study of the next generation sequencing of Globalfiler™ PCR amplification kit for the Ion Torrent S5 sequencer. Forensic Sci Int Genet Suppl Ser 7(1):336–338. https://doi.org/10.1016/j.fsigss.2019.10.002

Gettings KB, Borsuk LA, Ballard D, Bodner M, Budowle B, Devesse L, King J, Parson W, Phillips C, Vallone PM (2017) STRSeq: a catalog of sequence diversity at human identification short tandem repeat loci. Forensic Sci Int Genet 31:111–117. https://doi.org/10.1016/j.fsigen.2017.08.017

Guo F, Zhou Y, Song H, Zhao J, Shen H, Zhao B, Liu F, Jiang X (2016) Next generation sequencing of SNPs using the HID-Ion AmpliSeq™ Identity Panel on the Ion Torrent PGM™ platform. Forensic Sci Int Genet 25:73–84. https://doi.org/10.1016/j.fsigen.2016.07.021

Guo F, Yu J, Zhang L, Li J (2017) Massively parallel sequencing of forensic STRs and SNPs using the Illumina® ForenSeq™ DNA signature prep kit on the MiSeq FGx™ forensic genomics system. Forensic Sci Int Genet 31:135–148. https://doi.org/10.1016/j.fsigen.2017.09.003

Hoogenboom J, van der Gaag KJ, de Leeuw RH, Sijen T, de Knijff P, Laros JF (2017) FDSTools: a software package for analysis of massively parallel sequencing data with the ability to recognise and correct STR stutter and other PCR or sequencing noise. Forensic Sci Int Genet 27:27–40

Jäger AC, Alvarez ML, Davis CP, Guzmán E, Han Y, Way L, Walichiewicz P, Silva D, Pham N, Caves G, Bruand J, Schlesinger F, Pond SJK, Varlaro J, Stephens KM, Holt CL (2017) Developmental validation of the MiSeq FGx forensic genomics system for targeted next generation sequencing in forensic DNA casework and database laboratories. Forensic Sci Int Genet 28:52–70

King JL, Woerner AE, Mandape SN, Kapema KB, Moura-Neto RS, Silva R, Budowle B (2021) STRait Razor online: an enhanced user interface to facilitate interpretation of MPS data. Forensic Sci Int Genet 52:102463. https://doi.org/10.1016/j.fsigen.2021.102463

Meiklejohn KA, Robertson JM (2017) Evaluation of the precision ID Identity Panel for the Ion Torrent™ PGM™ sequencer. Forensic Sci Int Genet 31:48–56. https://doi.org/10.1016/j.fsigen.2017.08.009

Snedecor J, Fennell T, Stadick S, Homer N, Antunes J, Stephens K, Holt C (2022) Fast and accurate kinship estimation using sparse SNPs in relatively large database searches. Forensic Sci Int Genet 61:102769. https://doi.org/10.1016/j.fsigen.2022.102769

Stephens KM, Barta R, Fleming K, Perez JC, Wu SF, Snedecor J, Holt CL, LaRue B, Budowle B (2023) Developmental validation of the ForenSeq MainstAY kit, MiSeq FGx sequencing system and ForenSeq

Universal Analysis Software. Forensic Sci Int Genet 64:102851. https://doi.org/10.1016/j.fsigen.2023.102851

Tao R, Qi W, Chen C, Zhang J, Yang Z, Song W, Zhang S, Li C (2019) Pilot study for forensic evaluations of the precision ID GlobalFiler NGS STR panel v2 with the Ion S5 system. Forensic Sci Int Genet 43:102147

Zhou Y, Guo F, Yu J, Liu F, Zhao J, Shen H, Zhao B, Jia F, Sun Z, Song H, Jiang X (2016) Strategies for complete mitochondrial genome sequencing on Ion Torrent PGM™ platform in forensic sciences. Forensic Sci Int Genet 22:11–21. https://doi.org/10.1016/j.fsigen.2016.01.004

Statistical Interpretation of Forensic DNA Evidence

8

8.1 Introduction

The aim of forensic DNA analysis is individualization of any biological sample. In forensic science there is a concept called class characteristics and individual characteristics. The class characteristics are common to a group of objects or persons, whereas, the individual characteristics can be attributed to a particular individual. For example, blood group determination does not hold significance in individualization of forensic biological sample analysis. Blood group is not an individual characteristic, it is a class characteristic. 44.29% of the global indigenous populations share "O" blood group. Similarly, "A," "B," and "AB" blood groups are shared among 34.08%, 16.8%, and 5.21% of populations, respectively. Alternatively, it can be said that, if we determine B blood group in a crime scene exhibit, it is common to 1,344,000,000 people, considering the current world population to be eight billion. Other factors also come into play while estimating a match in forensic analysis. The most common biological parameter is the population-specific parameter. If we consider the Indian population, the B blood group is found to be most prevalent in around 31.68% individuals (Kumar et al. 2018). Considering this fact, if we determine "B" blood group in crime scene, the match probability becomes 1 in 3.15 or 449,856,000 people, considering the current Indian population to 1.42 billion.

Similarly, when we consider the forensic DNA analysis, it is an established fact that all humans share 99.9% of their DNA (Collins and Mansoura 2001). In forensic DNA analysis examinations, we do not focus on the 99.9% genome in common but the 0.1% of the genome that varies. Hence the genome of 0.1% non-coding region is analyzed. Out of ~three billion base pairs of total human genome, ~three million base pairs are of interest of a forensic DNA examiner. However, all of these ~three million base pairs are not examined in routine forensic DNA analysis practice. Only a relative few genetic markers are analyzed. As per the current recommendation, 20 expanded CODIS Short Tandem Repeat (STR) markers are analyzed. On average, one STR occurs in every 2000 bp of human genome and over half million STR markers have been characterized which constitutes around 3% of the entire human genome (Fan and Chu 2007; Tang and Nzabarushimana 2017). As a selective number of STR markers are explored in forensic DNA analysis, the chance of random match of the DNA profile in a population cannot be ruled out absolutely. Hence, the statistical evaluation of a STR DNA profile is carried out before making a conclusion.

Another fundamental aspect of forensic DNA analysis is "genome is inherited from the parents." Theoretically, each parent contributes half of their genome to their offspring. This fundamental aspect plays a vital role in establishing the identity of an individual by forensic DNA

H. R. Dash et al., *Advancements in Forensic DNA Analysis*,
https://doi.org/10.1007/978-981-99-6195-5_8

analysis. In cases of paternity dispute or identification of mutilated bodies, the DNA profile of the questioned sample is compared with the DNA profile obtained from the putative mother and the father. Mendelian inheritance of alleles is the basis of such identification. As we test only a few genetic markers, the chance of any random man in the population contributing allele to the questioned sample cannot be ruled out. Hence, statistical analysis is carried out for the match probability. Similarly, if the identification is made by comparing among the siblings, statistical interpretation of the putative relationship match is also carried out. In many sample limiting conditions, identification is carried out using the lineage markers, i.e., either Y-STRs or analysis of mitochondrial DNA.

Besides single source profiles, DNA profiles from mixed sources are generated in forensic DNA analysis. Samples involved in sexual assault cases and touch DNA samples are the common sources of mixed DNA profile generation. Due to the presence of multiple alleles in a genetic marker, interpretation of mixed profile becomes immensely difficult. The attribution of major contributor and minor contributor(s) in a mixed profile needs strong expertise. Multiple alleles further impact in the statical attribution of individual of a contributor's DNA profile to the mixed profile. Hence, different approaches of statistical evaluation of mixed DNA profile are carried out.

8.2 Two Fallacies

During interpretation of statistically evaluated forensic evidence, two fallacies commonly occur, i.e., the "prosecutor's fallacy" and the "defendant's fallacy." Both of these should be avoided. The "prosecutor's fallacy" is also called as the fallacy of the transposed conditional. It is popularly called as "getting the wrong end of a stick." This fallacy confuses the two conditional probabilities. For example, "P" is the probability of a match between a crime scene genotype and a suspect's genotype. The fallacy is to consider that, "P" is also the probability of the crime scene's genotype came from someone other than the suspect. An likelihood ratio (LR) of 100 states that the genotype match is 100 times as great if the evidence and the suspect samples that share the same profile are from the same person as it is if the samples are from different sources. It does not state that the suspect contributed the evidence DNA as 100:1. Such a probability is obtained by using Bayes's theorem and a conditional probability. The prior conditional probability is achieved on the basis of non-DNA evidences. If the prior probability is 1/2, the posterior odds equal the LR (Sesardic 2008).

In contrary, the defendant's fallacy assumes that anyone with the same DNA profile in the defined population is likely to have left the sample as the suspect. If ten persons exist with the same DNA profile as the evidence sample in an area, the defendant's fallacy is to assume that the probability that the suspect contributed the sample is only 0.1. However, the suspect was originally identified by other evidence, which is highly unlikely to match with the rest nine individuals having the same DNA profile (Sesardic 2008).

8.3 Choosing a Population-Specific Database

The DNA profile obtained from the crime scene can be considered to come from any random population whose database is available. The weight of evidence (WoE) can be evaluated by comparing the obtained DNA profile with the database. Hence, the choice of database becomes utmost importance while evaluating the WoE. The rarer an allele, the stronger the evidence becomes in implicating the suspect (Steele and Balding 2014). Though many population-specific allele frequency databases are present for the 13 CODIS markers, the 20 expanded CODIS marker database is still developing. Population databases to be used in forensic DNA analysis are created by compiling the genotypes of individuals from various ethnic or racial groups. Mostly, the population subdivisions are not considered while studying the distributing of alleles (Panneerchelvam and Norazmi 2003). It has been suggested that a sample size of greater than 100 is sufficient enough to predict the genotype frequency of VNTR markers in a larger population (Chakraborty 1992). However, the selected subjects for the creation of allele frequency database in a population should be randomly mating and unrelated individuals.

For analysis of the weight of DNA evidence, it is advocated to use the population-specific allele frequency databases. Though the allele frequency database for the microsatellite markers is well established for many developed countries, many countries still lack a population-specific database of their own. In this regard, a recent study showed no statistically significant differences in the allele frequency and other forensic and paternity parameters among different populations. 100 paternity trio cases without any inconsistencies were evaluated using different population databases and the cumulative paternity index (CPI) and probability of paternity (POP) did not show statistically significant variation when tested with the allele frequency value from different population databases. This study further suggests no urgent need for the use of population-specific allele frequency databases for the statistical evaluation of paternity trio cases with no inconsistencies (Dash et al. 2021).

8.4 Allele Frequency

Allele frequency can be defined as the relative frequency of alleles at a particular STR locus in a population. Each ethnic group has a specific allele frequency value. Hence, it may be possible to distinguish among different ethnic groups based on their allele frequency values. In forensic DNA analysis, allele frequency is the precursor for the statistical evaluation of the forensic DNA evidences by calculating various parameters such as random match probability, likelihood ratios, and mixture interpretation (Park et al. 2021). Allele frequency can be calculated by counting the number of occurrences of each allele in all the samples. In many instances, the allele frequency value does not reflect its actual nature, especially in cases of a very rare allele because the rare allele is not represented in a sufficient number of instances during the observation. Hence, it has been recommended that each allele should be observed at least five times to be used in forensic statistical calculations. Hence, the minimum value of an allele frequency is $5/(2n)$, where "n" is the number of individuals included in the study and "$2n$" is the number of total chromosomes (Butler et al. 2009).

Various authors have used different nomenclatures to describe alleles, genotypes, and allele frequencies. Use of such varied nomenclature may cause confusion among readers to interpret the results of allele frequency. The commonly used abbreviations for alleles, genotypes, and allele frequencies in literature is given in Table 8.1.

8.5 Genotype Frequency

Genotypes are of two types, i.e., homozygous and heterozygous. Using the allele frequency value, the genotype frequency is calculated. Diploid individuals have two autosomal genes, one from each parent. When the allele inherited from sperm and egg differ, an person is heterozygous for that locus. Similarly, when an individual receives two identical alleles from both the parents it constitutes homozygous genotype. As the number of possible alleles in a genetic marker increase, the number of possible genotypes also increase subsequently. For a large number of alleles as observed in case of the DNA markers, the number of possible genotypes is significantly high. Generally, if a genetic marker has "n" possible alleles, it may produce "n" possible homozygous genotypes and "$n(n - 1)/2$" possible heterozygous genotypes. Hence, in total the number of possible genotypes for "n" no. of alleles is "$n(n + 1)/2$." For examples, if a genetic marker has ten possible alleles in a population, it will exhibit ten possible homozygous genotypes and $[10 \times (10-1)]/2$, i.e., 45 possible heterozygous

Table 8.1 Commonly used nomenclature for alleles, genotypes, and allele frequencies

Alleles	Genotype	Allele frequency
P, Q	PQ	p, q
$\boldsymbol{p, q}$	$\boldsymbol{pq}$	p, q
A_u, A_v	A_uA_v	p_u, p_v
A_i, A_j	A_iA_j	p_i, p_j
A_i, A_j	A_iA_j	p_i, p_j
A_1, A_2	A_1A_2	p_1, p_2
A, B	AB	Pr_A, Pr_B
a, b	ab	p_a, p_b
A, B	AB	P_A, P_B
a, b	ab	f_a, f_b

genotypes. In total there are 10 + 45 = 55 possible genotypes.

The allele frequencies follow Hardy-Weinberg equilibrium. The principle of Hardy-Weinberg equilibrium is that allele frequencies are inherited in a Mendelian fashion and frequencies of occurrence follow a predictable pattern of probability. If there are two alleles in a genetic marker, i.e., "*P*" and "*Q*" with their allele frequencies "*p*" and "*q*," the frequency of homozygous genotypes is "*pp*" and "*qq*." Similarly, the frequency of heterozygous genotype is "2*pq*." As per Hardy-Weinberg principle, $(p + q) = 1$ and subsequently the sum of all genotype frequencies is also 1. It can be denoted as $p^2 + 2pq + q^2 = 1$. Such a frequency of alleles and genotypes will remain unchanged in generations unless other evolutionary forces come into play. Genotype frequency can be determined by calculating the number of individuals with a particular genotype divided by the total number of individuals examined. The relationship between the allele frequency and genotype frequency is shown in Fig. 8.1.

8.6 Random Match Probability

The weight of a single source DNA profile is mostly assessed by calculating the random match probability (RMP). The calculated allele frequency and genotype frequency values are used for calculating RMP of a DNA profile. RMP signifies how rare the DNA profile is. It determines the probability of a random person in the population having the same DNA profile. The calculating of RMP is based on the product rule of probability. The product rule of probability is applicable for the phenomenon of independent transmission of characteristics. This rule states that the probability of two independent events occurring simultaneously can be calculated by multiplying the individual probabilities of each event alone.

In routine forensic DNA analysis, the selected STR markers follow Mendelian inheritance pattern. None of these markers also show linkage disequilibrium. Hence, all the alleles present in the STR markers show independent assortment during the formation of zygote. Hence, when a set of STR markers are evaluated in a DNA profile, RMP can be calculated by using the product rule of probability. The example of such a calculation of RMP is given in Table 8.2.

8.7 Paternity Index

Parentage testing in forensic DNA analysis is based on the principle that half of the genome of a child is inherited from each of the biological parents. A set of genetic markers are analyzed and matched among the DNA profile obtained from the mother, child, and the alleged father. Two possible outcomes are possible. If the alleles do not match at three or more STR loci, the test result is regarded as exclusion. In this regard, a study shows that 5 to 18 STR markers are responsible for the exclusion in most of the paternity

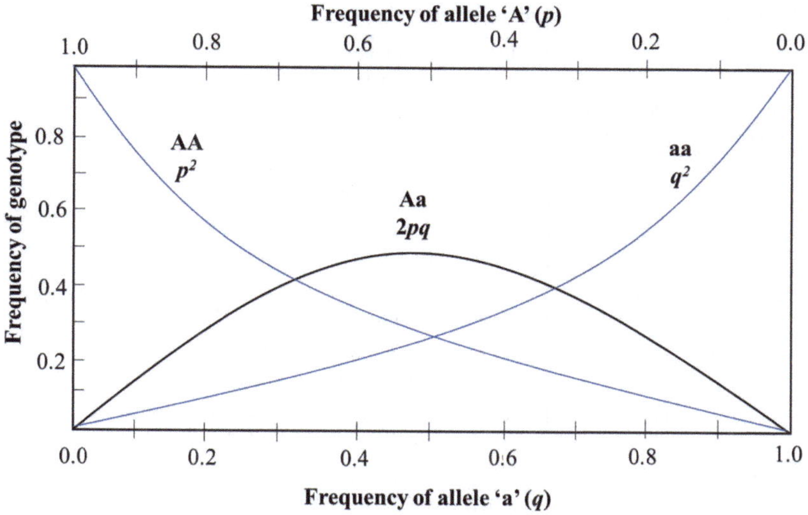

Fig. 8.1 The relationship between the allele frequency and genotype frequency under Hardy-Weinberg equilibrium condition

exclusion cases. In case of motherless paternity, the number of STR markers responsible for exclusion increases further to 14 numbers (Gonzalez et al. 2019). Such exclusion results do not need a statistical evaluation of the result. However, when the obligate paternal alleles in the child is found in the corresponding STR marker of the alleged father, it is called inclusion and a statistical evaluation of the weightage of such evidence is required.

Paternity Index (PI) is calculated in all cases of paternity inclusions. PI is calculated using the following formula:

$$\mathrm{PI} = X/Y$$

where

X = the probability that alleged father is the biological father

Y = the probability that any random man in the population is the biological father

PI is calculated in all the STR markers tested individually followed by the calculation of combined paternity index (CPI) by using the product rule of probability. CPI signifies the chance of any random man unrelated man of the population to be the biological father of the child. Using the CPI value, the probability of paternity is calculated using the formula:

$$\text{Probability of Paternity} = \mathrm{CPI}/(1 + \mathrm{CPI}) \times 100$$

8.7.1 Calculation of PI in Paternity Trio

PI is calculated in a paternity trio case using the STR profiles of child, mother, and the father. A representative calculation of PI in five STR markers have been provided here (Table 8.3). However, such calculations can be extended up to the number of STR markers tested. Another prerequisite for the calculation of PI is the allele frequency value of the STR markers in the population. Such allele frequency values can be taken from the population database or a study can be performed to generate the allele frequency values as described in the previous section.

Using the three sets of alleles of the STR markers, obligate alleles are calculated. In paternity trio cases in forensic DNA analysis mother's sample is considered to be known. After comparing the allelic data of the mother and the child at a particular STR marker, "the allele which must be present in the father" is predicted and is called as the obligate allele. Obligate paternal allele can be defined as the presence of this particular allele in the father's corresponding genetic marker in case of a deemed paternity inclusion. In contrary, if three or more of the obligate allele are not found

Table 8.2 Calculation of random match probability (RMP) of a single source DNA profile

Genetic marker	Observed genotype	Allele frequency in the database	Formula to calculate genotype frequency	Calculated genotype frequency	RMP of the DNA profile
D3S1358	(14,17)	14: 0.049 17: 0.237	$2 \times p \times q$ (heterozygous)	$2 \times 0.049 \times 0.237 = 0.0232$	$0.0232 \times 0.0334 \times 0.1188 \times 0.2301 \times 0.1673 \times 0.0259 \times 0.0511 \times 0.0177 \times 0.0805 \times 0.0085 \times 0.1538 \times 0.0191 \times 0.0449 \times 0.1011 \times 0.0323 \times 0.0137 \times 0.0007 \times 0.1545 \times 0.0653 \times 0.0124 \times 0.0580 =$ **1.7024×10^{-30}**
vWA	(15,17)	15: 0.068 17: 0.246	$2 \times p \times q$ (heterozygous)	$2 \times 0.068 \times 0.246 = 0.0334$	
D16S539	(11,12)	11: 0.330 12: 0.180	$2 \times p \times q$ (heterozygous)	$2 \times 0.330 \times 0.180 = 0.1188$	
CSF1PO	(11,12)	11: 0.282 12: 0.408	$2 \times p \times q$ (heterozygous)	$2 \times 0.282 \times 0.408 = 0.2301$	
TPOX	(11,11)	11: 0.409	$p \times p$ or $q \times q$ (homozygous)	$0.409 \times 0.409 = 0.1673$	
D8S1179	(10,12)	10: 0.180 12: 0.072	$2 \times p \times q$ (heterozygous)	$2 \times 0.180 \times 0.072 = 0.0259$	
D21S11	(28,30)	28: 0.126 30: 0.203	$2 \times p \times q$ (heterozygous)	$2 \times 0.126 \times 0.203 = 0.0511$	
D18S51	(13,17)	13: 0.132 17: 0.067	$2 \times p \times q$ (heterozygous)	$2 \times 0.132 \times 0.067 = 0.0177$	
D2S441	(11,14)	11: 0.376 14: 0.107	$2 \times p \times q$ (heterozygous)	$2 \times 0.376 \times 0.107 = 0.0805$	
D19S433	(15,16)	15: 0.095 16: 0.045	$2 \times p \times q$ (heterozygous)	$2 \times 0.095 \times 0.045 = 0.0085$	
TH01	(6,9)	6: 0.209 9: 0.368	$2 \times p \times q$ (heterozygous)	$2 \times 0.209 \times 0.368 = 0.1538$	
FGA	(20,25)	20: 0.076 25: 0.126	$2 \times p \times q$ (heterozygous)	$2 \times 0.076 \times 0.126 = 0.0191$	
D22S1045	(11,17)	11: 0.330 17: 0.068	$2 \times p \times q$ (heterozygous)	$2 \times 0.330 \times 0.068 = 0.0449$	
D5S818	(12,12)	12: 0.318	$p \times p$ or $q \times q$ (homozygous)	$0.318 \times 0.318 = 0.1011$	
D13S317	(12,13)	12: 0.207 13: 0.078	$2 \times p \times q$ (heterozygous)	$2 \times 0.207 \times 0.078 = 0.0323$	
D7S820	(7,10)	7: 0.032 10: 0.214	$2 \times p \times q$ (heterozygous)	$2 \times 0.032 \times 0.214 = 0.0137$	
SE33	(17,22)	17: 0.045 22: 0.008	$2 \times p \times q$ (heterozygous)	$2 \times 0.045 \times 0.008 = 0.0007$	
D10S1248	(14,15)	14: 0.281 15: 0.275	$2 \times p \times q$ (heterozygous)	$2 \times 0.281 \times 0.275 = 0.1545$	
D1S1656	(11,15)	11: 0.191 15: 0.171	$2 \times p \times q$ (heterozygous)	$2 \times 0.191 \times 0.171 = 0.0653$	
D12S391	(21,22)	21: 0.086 22: 0.072	$2 \times p \times q$ (heterozygous)	$2 \times 0.086 \times 0.072 = 0.0124$	
D2S1338	(19,23)	19: 0.215 23: 0.135	$2 \times p \times q$ (heterozygous)	$2 \times 0.215 \times 0.135 = 0.0580$	

in the father, it becomes a case of exclusion. Obligate allele can be one or two in numbers. The predicted obligate allele in the abovementioned case is given in Table 8.4.

Table 8.3 Allelic data of father, mother and child at five STR markers

Genetic marker	Mother	Child	Father
D3S1358	14,17	17,19	18,19
vWA	11,12	11,12	11,13
D16S539	12,13	12,13	12,13
CSF1PO	10,10	10,10	10,10
TPOX	10,11	11,12	12,12

Table 8.4 Prediction of obligate allele

Genetic marker	Mother	Child	Obligate allele
D3S1358	14,17	17,19	19
vWA	11,12	11,12	11,12
D16S539	12,13	12,13	12,13
CSF1PO	10,10	10,10	10
TPOX	10,11	11,12	12

After predicting the obligate allele (OA), the allele frequency of each OA is determined from the population database. Furthermore, the likelihood that the alleged father could transmit the OA is calculated which is denoted as "*X*." The value of *X* is either 0.5 or 1.0. The thumb rule of the likelihood prediction is that, if the father is homozygous at that particular STR marker, "*X*" becomes 1.0 and in case the father is heterozygous, the value of "*X*" becomes 0.5. Using the allele frequency and likelihood values, the PI is calculated using the formula: PI = Likelihood (*X*)/Allele Frequency. When there are two obligate alleles, the sum of both the allele frequencies is taken in the denominator. Finally, the combined paternity index (CPI) is calculated by multiplying the individual PIs of all the genetic markers tested (Table 8.5). Using CPI value, probability of paternity can be calculated using the formula PP = CPI/(1 + CPI)x100. So for the calculated CPI value in Table 8.5, the computed PP is 98.743%.

8.7.2 Calculation of PI in Paternity Duo

In real-time forensic DNA cases, sample limiting conditions arise more often. Motherless paternity cases can be considered as the paternity duo case, where PI is calculated by matching only two DNA profiles, the DNA profile of the child and the alleged father. In such cases, the paternity index can be calculated using the formula given in Table 8.6.

8.7.3 Sibship Index

Sibship index (SI) is calculated to evaluate a match and to assign a relationship as full or half siblings. A combined half sibship index of more than one is generally regarded as the evidence in favor of the tested relationship. Unlike in paternity assessment, there is no obligatory paternal allele from where a person can be excluded logically. In case of true siblings except the homozygous twins, three possible observations are found theoretically, i.e., two shared alleles, one shared allele and no shared allele with a frequency of 0.25, 0.50, and 0.25, respectively. Siblings as well as other unrelated persons can also share these alleles, mostly when these alleles are most common in the population. SI is calculated as the first conditional probability of the tested individual is the sibling of the questioned individual. The second conditional probability is that any random man in the population is the sibling of the questioned individual.

For calculation of sibship index, the kinship coefficient (*k*) and population allele frequency value are required. When S1 (sibling 1) and S2

Table 8.5 Calculation of likelihood, paternity index, and combined paternity index in a paternity trio case

Genetic marker	Mother	Child	Obligate allele	Frequency of OA	Father	*X*	Formula for PI	PI	Combined PI
D3S1358	14,17	17,19	19	0.315	18,19	0.5	*X*/frequency	0.5/0.315 = 1.587	78.586
vWA	11,12	11,12	11,12	0.076, 0.29	11,13	0.5	*X*/frequency 1 + frequency 2	0.5/(0.076 + 0.29) = 1.366	
D16S539	12,13	12,13	12,13	0.188, 0.035	12,13	0.5	*X*/frequency 1 + frequency 2	0.5/(0.188 + 0.035) = 2.242	
CSF1PO	10,10	10,10	10	0.159	10,10	1.0	*X*/frequency	1.0/0.159 = 6.289	
TPOX	10,11	11,12	12	0.389	12,12	1.0	*X*/frequency	1.0/0.389 = 2.571	

Table 8.6 Formula for calculating PI in paternity duo cases while mother's sample is not available (adopted from Chakraborty et al. 1994)

Genotype				
Child	Father	Truly related	Unrelated	PI
AB	AC	abc	$4a^2bc$	$1/4a$
AB	AB	$ab(a + b)$	$(2ab)^2$	$(1/4a) + (1/4b)$

Table 8.7 Possible paired genotypes and their corresponding k coefficients (adopted from Wenk et al. 1996)

Genotype		Probability of S2 genotype if two persons		
S1	S2	Are truly related (R)	Are unrelated (U)	SI = RU
Full siblings				
AA	AA	$1(k_2) + a\ (k_1) + a\ (k_1) + a^2\ (k_0)$	a^2	$k_2 + 2k_1a + k_0a^2/a^2$
AA	AB	$b(k_1) + b(k_1) + 2ab(k_0)$	$2ab$	$(k_1 + k_0a)/a$
AA	BB	$b^2(k_0)$	b^2	k_0
AA	BC	$2bc(k_0)$	$2bc$	k_0
AB	AB	$1(k_2) + a(k_1) + b(k_1) + 2ab(k_0)$	$2ab$	$(k_2 + k_1a + k_1b + k_02ab)/2ab$
AB	AA	$a(k_1) + a^2(k_0)$	a^2	$(k_1 + k_0a)/a$
AB	AC	$c(k_1) + 2\ ac(k_0)$	$2\ ac$	$(k_1 + 2k_0a)/2a$
AB	CD	$2\ cd(k_0)$	$2\ cd$	k_0
Half siblings				
AA	AA	$a(k_1) + a(k_1) + a^2(k_0)$	a^2	$(2k_1 + k_0a)/a$
AA	AB	$k_1(b) + k_1(b) + 2ab\ (k_0)$	$2ab$	$(k_1 + k_0a)/a$
AA	BB	$b^2(k_0)$	b^2	k_0
AA	BC	$2bc(k_0)$	$2bc$	k_0
AB	AB	$a(k_1) + b\ (k_1) + 2ab\ (k_0)$	$2ab$	$(k_1a + k_1b + k_02ab)/2ab$
AB	AA	$a(k_1) + a^2(k_0)$	a^2	$(k_1 + k_0a)/a$
AB	AC	$c(k_1) + 2\ ac(k_0)$	$2\ ac$	$(k_1 + 2k_0a)/2a$
AB	CD	$2\ cd(k_0)$	$2\ cd$	k_0

(sibling 2) are tested, the established kinship coefficients are as follows:

k_2 = probability that both the alleles of the siblings are identical by descent
k_1 = one random allele is identical in both the tested individuals as identical by descent, whereas the second allele is not identical
$2k_1$ = probability that one or the other of the alleles in S2 is identical by descent to one of the alleles of S1, but the second is not identical
k_0 = probability that neither of the alleles of S1 is identical with the alleles of S2

For siblings, the established values of kinship coefficients are $k_2 = 0.25$, $k_1 = 0.25$, and $k_0 = 0.25$. For half siblings, $k_2 = 0$, $k_1 = 0.25$, and $k_0 = 0.50$. As the k probability describes all the possibilities of two relatives sharing the alleles among themselves, hence, $(k_2 + 2k_1 + k_0) = 1.0$. The various possibilities of k values in a family are given in Table 8.7. Calculation of k values is highly important in claiming kinship, when pedigree/genotype information is limited to only two individuals.

8.8 Avuncular Index

By calculating avuncular index (AI), the likelihood ratio of two competing hypothesis can be tested, i.e., H1: The tested man's brother is the biological father of the child and H2: A random man is the biological father of the child. The AI can be defined as the likelihood ratio which tests the hypothesis that the tested man is a paternal

Table 8.8 Modified formula to calculate PI in tri-allelic child (adopted from Yang et al. 2021)

Mother	Child	Putative Father	PI value
Type I tri-allelic pattern			
PP	P*QQ′	QR/Q′R	1/2(PQ + PQ′)
PP	P*QQ′	QQ/Q′Q′/QQ′	1/(PQ + PQ′)
QQ/Q′Q′	P*QQ′	PR	1/2PP
QQ/Q′Q′	P*QQ′	PP	1/PP
PQ	P*RR′	RS/R′S	1/2(PR + PR′)
PQ	P*RR′	RR/R′R′/RR′	1/(PR + PR′)
QQ′	P*QQ′	PR	1/2PP
QQ′	P*QQ′	PP	1/PP
PQ/PQ′	P*QQ′	QQ/Q′Q′/QQ′	1/(PQ + PQ′ + PP)
PQ/P′Q	P*QQ′	PQ/P′Q	1/(PQ + PQ′ + PP)
Type II tri-allelic pattern			
PP	PQR	QRS	PS/[2(PQ + PR + PS)i]
PQ	PQR	PQR	(PP + PQ)
PQ	QRR	RS	10−6PQ/2i(PQ + PR + PR)
PQ	QRR	RR	10−6PP/i(PP + PQ + PQ)
PQ	PQQ	PQ	10−6PP/2i(PP + PQ + PQ)
PQ	PQQ	QQ	10−6PP/2i(PP + PQ + PQ)

* Indicates the major allele in the Type I tri-allelic genotype

uncle of the child versus the hypothesis that the tested man is unrelated to the child. Hence, AI can be calculated as $(X + Y)/2$ = (PI+1)/2 (Morris et al. 1988).

8.9 Calculation of PI in Atypical Cases

In many instances, certain atypical DNA profiles are obtained while conducting paternity analysis. One such common atypical match involves the inconsistencies at one or two genetic markers out of all the STR markers analyzed. Such inconsistencies at one or two genetic markers are attributed to mutations rather than mismatch. In such cases, the formula to calculate is modified by incorporating the mutation rate of that particular STR marker. As per the recommendations of paternity testing commission (PTC) of the International Society of Forensic Genetics (ISFG), for inconsistency loci, the formula for PI has been modified as $\mathrm{PI} = \mu/\bar{A}$, where μ is the locus specific mutation rate and $\bar{A}$ is the average probability of exclusion (Di Lonardo et al. 2004).

Besides inconsistencies in few STR markers, observance of tri-allelic pattern is also often observed which complicates the calculation of Paternity Index. Two types of tri-allelic patterns are commonly observed. Type I tri-allelic pattern results from the somatic mutation observed during early stage of development. Whereas Type II tri-allelic pattern is generated by segmental duplication in the genome. The formula to calculate PI in various tri-allelic patterns is given in Table 8.8.

8.10 Haplotype Frequency

While analyzing the lineage markers found on Y chromosome or mitochondrial genome, haplotype frequency is calculated in most of the cases. When a pool of DNA contains samples from M number of subjects, a total of 2 M haplotype copies should be present in the pool. Out of 2 M number of haplotypes, some of the haplotypes may be repeated. Hence, haplotype frequencies can be calculated for each unique haplotype. The haplotype frequency can be calculated either manually or by using different computer programs such as, HapYDive (Alves et al.

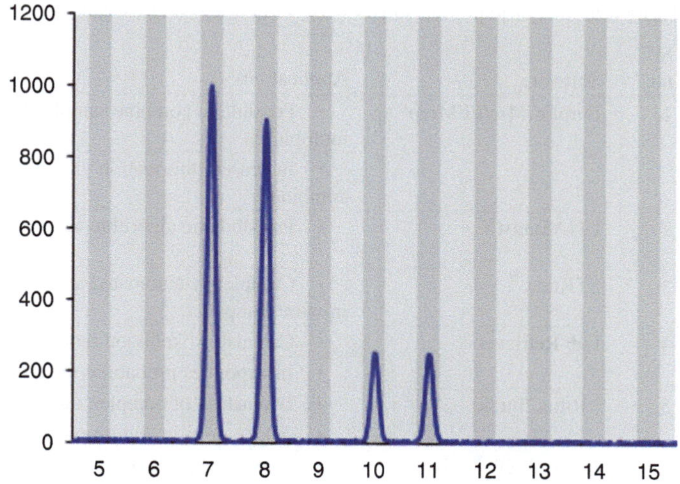

Fig. 8.2 A typical STR locus with more than one contributor

2006), ySTRmanager (Yang et al. 2012), or YHRD (Roewer 2003).

8.11 Interpretation of Mixed Profile

A mixed DNA profile is one which comprises DNA from more than one individual. The presence of mixed DNA profile complicates the analysis process in comparison to a single source DNA profile. The presence of three or more peaks at a given STR locus is a clear indication of a mixed source of DNA. In mixed profile interpretation, mere presence of alleles of the person of interest does not solve the purpose. DNA profile of a person of interest should be a good fit in the mixed profile also. For example, a representative mixed autosomal STR profile is shown in Fig. 8.2. Here, we can assume that two people having genotype (7,8) and (10,11) are the contributors to this DNA profile from the peak height ratios. Likewise, a person with genotype (7,11) or (8,10) is not a good fit to this DNA profile.

The first step of statistical analysis of a mixed profile is to determine the number of contributors. Determination of the number of contributors is subjective and complex process. Mostly the number of contributors is determined by looking at the locus with the most number of alleles. However, the presence of ≤ 2 peaks can also represent a DNA mixture with homozygotes. The statistical process used for the mixed DNA profiles is different than that of the statistical process used for the single source DNA profile. In the determination of the number of contributors of a DNA profile, this maximum allele count strategy may fail at some instances if the minor alleles are not considered properly. Many a times, the forensic examiner has a prior knowledge of the number of contributors to a mixed DNA profile in the sexual assault cases. In sexual assault cases, the presence of multiple peaks should be carefully observed to avoid error. In such cases, the presence of low levels of DNA from a consensual partner may be overlooked using modern techniques in 3% of such cases (Buckleton et al. 2007).

Calculation of the combined probability of inclusion (CPI) is the most widely used method to ascertain the weight of an evidence when a probative profile is obtained from an evidentiary sample. CPI denotes the proportion of a population that would be expected to be included as a potential contributor in an observed DNA mixture. Three steps are involved in calculation of CPI in a mixed profile, i.e., profile assessment, comparison with the mixed profile for determination of inclusion/exclusion and statistical

Table 8.9 List of important software useful in forensic DNA analysis

Sl. no.	Software	Applications	Reference
1	EuroForMix/EFMrep	• Possible to combine samples from different multiplexes • Increased information gain when samples are combined	Bleka et al. (2022)
2	DNAStatistX	• Probabilistic algorithm to calculate likelihood ratios	Benschop et al. (2019)
3	STRmix™	• Completely automated assessment of reference and mixture samples	Bright et al. (2019)
4	Lab Retriever	• Calculates likelihood ratio • Incorporates probability of dropout	Inman et al. (2015)
5	LRmix Studio	• Evaluation of complex mixture • Allelic drop-in and drop-out included	Gill and Haned (2013)
6	DNA-VIEW® mixture solution	• Combined probability of inclusion (CPI) • Mixture solution models	Lucassen et al. (2021)
7	LikeLTD	• Likelihood ratio (LR) • Useful in LTDNA cases	Steele et al. (2014)
8	True Allele	• Likelihood ratio (LR) • Validated for DNA mixture of up to 10 contributors	Bauer et al. (2020)

calculations (Bieber et al. 2016). Likelihood ratio calculation is also used in many cases involving mixed profile interpretation.

Now-a-days many probabilistic genotyping software programs are available to evaluate the forensic DNA evidence and to interpret the mixed profiles as well as other applications. Some of this important software is listed in Table 8.9. The next chapter will focus on databases.

References

Alves C, Gusmao L, Meirinhos J, Amorim A (2006) Making the most of Y-STR haplotypes. HapYDive Progr Forensic Genet 11:201–203

Bauer DW, Butt N, Hornyak JM, Perlin MW (2020) Validating TrueAllele® interpretation of DNA mixtures containing up to ten unknown contributors. J Forensic Sci 65:380–398. https://doi.org/10.1111/1556-4029.14204

Benschop CCG, Hoogenboom J, Hovers P, Slagter M, Kruise D, Parag R, Steensma K, Slooten K, Nagel JHA, Dieltjes P, van Marion V, van Paassen H, de Jong J, Creeten C, Sijen T, Kneppers ALJ (2019) DNAxs/DNAStatistX: development and validation of a software suite for the data management and probabilistic interpretation of DNA profiles. Forensic Sci Int Genet 42:81–89. https://doi.org/10.1016/j.fsigen.2019.06.015

Bieber FR, Buckleton JS, Budowle B, Butler JM, Coble MD (2016) Evaluation of forensic DNA mixture evidence: protocol for evaluation, interpretation, and statistical calculations using the combined probability of inclusion. BMC Genet 17:125. https://doi.org/10.1186/s12863-016-0429-7

Bleka O, Prieto L, Gill P (2022) EFMrep: an extension of EuroForMix for improved combination of STR DNA mixture profiles. Forensic Sci Int Genet 61:102771. doi.org/10.1016/j.fsigen.2022.102771

Bright JA, Cheng K, Kerr Z, McGovern C, Kelly H, Moretti TR, Smith MA, Bieber FR, Budowle B, Coble MD, Alghafri R, Allen PS, Barber A, Beamer V, Buettner C, Russell M, Gehrig C, Hicks T, Charak J, Wing KC, Buckleton J (2019) STRmix™ collaborative exercise on DNA mixture interpretation. Forensic Sci Int Genet 40:1–8. https://doi.org/10.1016/j.fsigen.2019.01.006

Buckleton JS, Curran JM, Gill P (2007) Towards understanding the effect of uncertainty in the number of contributors to DNA stains. Forensic Sci Int Genet 1: 20–28

Butler JM, Buel E, Crivellente F, McCord B (2009) Forensic DNA typing by capillary electrophoresis using the ABI Prism 310 and 3100 genetic analyzers for STR analysis. Electrophoresis 25:1397–1412. https://doi.org/10.1002/elps.200305822

Chakraborty R (1992) Sample size requirements for addressing the population genetic issues of forensic use of DNA typing. Hum Biol 642:141–159

Chakraborty R, Jin L, Zhong Y (1994) Paternity evaluation in cases lacking a mother and nondetectable alleles. Int J Legal Med 107:127–131. https://doi.org/10.1007/BF01225599

Collins FS, Mansoura MK (2001) The human genome project. Revealing the shared inheritance of all humankind. Cancer 91:221–225. https://doi.org/10.1002/1097-0142

Dash HR, Vajpayee K, Agarwal R, Gang A, Shukla R, Srivastava A (2021) Evaluation of diallelic STR markers with inter-population allelic database for their usefulness in paternity trios in the Central Indian population. Ann Hum Biol 48:605–613. https://doi.org/10.1080/03014460.2021.2014567

Di Lonardo AM, Rodrıguez Cardozo MB, Colica MV, Abovich MA, Echenique C (2004) Genetic inconsistencies in paternity cases. Int Congr Ser 1261:488–490

Fan H, Chu JY (2007) A brief review of short tandem repeat mutation. Genomics Proteomics Bioinformatics 5:7–14. https://doi.org/10.1016/S1672-0229(07)60009-6

Gill P, Haned H (2013) A new methodological framework to interpret complex DNA profiles using likelihood ratios. Forensic Sci Int Genet 7:251–263. https://doi.org/10.1016/j.fsigen.2012.11.002

Gonzalez PNL, Bautista-Gonzalez JE, Lopez-Gonzalez MJ, Sosa-Escalante JE, Gonzalez-Herrera L (2019) The most frequent autosomal STRs involved in exclusion of paternity cases in a population from southeast, Mexico. Forensic Sci Int Genet Suppl Ser 7:465–467. https://doi.org/10.1016/j.fsigss.2019.10.053

Inman K, Rudin N, Cheng K, Robinson C, Kirschner A, Inman-Semerau L, Lohmueller KE (2015) Lab retriever: a software tool for calculating likelihood ratios incorporating a probability of drop-out for forensic DNA profiles. BMC Bioinformatics 16:298. https://doi.org/10.1186/s12859-015-0740-8

Kumar S, Modak PK, Ali SH, Barpanda SK, Gusain VS, Roy R (2018) ABO and Rh phenotype blood group distribution among blood donors in H.N.B. Base Hospital, Srinagar, Uttarakhand, India. J Family Med Prim Care 7:34–38. https://doi.org/10.4103/jfmpc.jfmpc_252_17

Lucassen A, Ehlers K, Grobler PJ, Brenner CH (2021) Evaluating Mixture Solution™-rapid and non-MCMC probabilistic mixture analysis. Int J Legal Med 135:2275–2284. https://doi.org/10.1007/s00414-021-02680-2

Morris JW, Garber RA, D'Autremont J, Brenner CH (1988) The avuncular index and the incest index. In: Mayr WR (ed) Advances in forensic haemogenetics, vol 2. Springer, Berlin. https://doi.org/10.1007/978-3-642-73330-7_124

Panneerchelvam S, Norazmi MN (2003) Forensic DNA profiling and database. Malaysian J Med Sci 10:20–26

Park HC, Ahn ER, Shin SC (2021) Comparative analysis of allele variation using allele frequencies according to sample size in Korean population. Genes Genomics 43:1301–1305. https://doi.org/10.1007/s13258-021-01159-z

Roewer L (2003) The Y-short tandem repeat haplotype reference database (YHRD) and male population stratification in Europe—impact on forensic genetics. Forensic Sci Rev 15(2):165–172

Sesardic N (2008) Guilt by statistical association: revisiting the prosecutor's fallacy and the interrogator's fallacy. J Philos 105:320–322. https://doi.org/10.5840/jphil2008105640

Steele CD, Balding DJ (2014) Choice of population database for forensic DNA profile analysis. Sci Justice 54:487–493. https://doi.org/10.1016/j.scijus.2014.10.004

Steele CD, Greenhalgh M, Balding DJ (2014) Verifying likelihoods for low template DNA profiles using multiple replicates. Forensic Sci Int Genet 13:82–89. https://doi.org/10.1016/j.fsigen.2014.06.018

Tang H, Nzabarushimana E (2017) STRScan: targeted profiling of short tandem repeats in whole-genome sequencing data. BMC Bioinformatics 18:398. https://doi.org/10.1186/s12859-017-1800-z

Wenk RE, Traver M, Chiafari FA (1996) Determination of sibship in any two persons. Transfusion 36:259–262

Yang IS, Lee HY, Park MJ, Yang WI, Shin KJ (2012) Web-based Y-STR database for haplotype frequency estimation and kinship index calculation. Korean J Legal Med 36:45–55. https://doi.org/10.7580/KoreanJLegMed.2012.36.1.45

Yang Q, Yao Y, Shao C, Zhou Y, Li H, Li C, Tang Q, Xie J (2021) Calculation of the paternity index for STR with tri-allelic patterns in paternity testing. Forensic Sci Int 324:110832. https://doi.org/10.1016/j.forsciint.2021.110832

9 Role of Forensic DNA Databases in Criminal Identification

9.1 Introduction

Over the past 20 years, the creation of national forensic DNA databases has revolutionized individual identification in the criminal justice system (Teodorović et al. 2017). There is no question that the existence of DNA criminal databases enhances the control and investigation of crime in the field of forensic genetics. The field of forensic genetics applies genetic principles and methods to law enforcement and the legal system. DNA samples are currently collected from crime scenes, suspects (which suspects and under what conditions vary by country), and "volunteers" for the purpose of elimination. Samples and profiles may be permanently or temporarily preserved, routinely checked for matches with crime scene samples, utilized for ancestry searches, and subjected to research pertinent to crime detection, depending on the laws of the country in question. With the help of modern DNA analysis methods like "low copy number" (LCN) testing, a profile can be created from smaller and lower quality DNA samples.

In 2008, the Greek government passed legislation that established and governed the National DNA Database of Greece. Only DNA profiles of convicts with convictions were recorded under this law. But a year later, in 2009, the law was changed to allow for the expansion of the database to include innocent people and minors. Because such decisions may have an impact on the entire community, dealing with databases and their control depends on what local communities are ready to allow (Ortner and Whitehead 1992).

In terms of forensic DNA databases, there are currently three options accessible, each with pros and cons. The first option—the laxest—is based on genotyping all and includes the general population in the database. The second, a more conservative option, only allows DNA fingerprinting for a particular list of crimes and the inclusion of profiles in the database for people who are strongly associated with those crimes. The third is the last, and it opposes the creation of DNA databases for forensic analysis (Guillén et al. 2000).

This second method states that the individual will only be ordered to undergo testing under duress if there is conclusive evidence connecting them to the crime. Nevertheless, this strategy ensures that the DNA analysis results won't be utilized for anything other than what they were originally intended for because they aren't kept in a database.

According to the Council for Responsible Genetics (2011), 26 countries are currently planning DNA registers, while 56 maintain functioning DNA databases (Thibedeau 2011). However, due to differences in legislation, sample collection, inclusion and removal criteria, profile and sample retention, database access, etc., different countries run their DNA databases in different ways, leading to a range of methods, from restrictive to permissive (Van Camp and Dierickx 2008). Due to the above mentioned, international

H. R. Dash et al., *Advancements in Forensic DNA Analysis*,
https://doi.org/10.1007/978-981-99-6195-5_9

studies have shown that the use of forensic DNA databases creates a number of issues, especially considering the seriousness of the outcomes, which can result in convictions and exonerations of people (Levitt 2007).

The negative effects on personal privacy and civil rights, potential racial and medical discrimination, public mistrust of the government (because to the fraudulent use of DNA in criminal investigations), information sharing with third parties, and genetic surveillance are all mentioned by opponents. Therefore, responsible genetics is essential for guaranteeing a higher degree of security, the protection of human rights and the individual freedoms of all people, and striking a balance between the utility and convenience of DNA data in law enforcement organizations (Williams and Johnson 2005). There are neither moral nor legal problems with comparing forensic profiles since criminals who dispose of their DNA at crime sites are not believed to have a right to privacy. Expanded DNA profiling on unresolved cases is highly supported since forensic laboratories are mandated to maximize the value of evidence and law enforcement is required to solve crimes using all available legal and ethical options. Due to the fact that criminals are still at large and might still conduct more crimes before being apprehended, solving unsolved crimes helps prevent the victims of future crimes and makes communities safer.

9.2 Custody of the Databases

The type of institution that should have custody of the biological samples obtained and the DNA profiles produced is a crucial topic that has caused debate in several nations. This is very important because the public's trust in DNA databases depends on their transparency and accountability (Voultsos et al. 2011).

9.3 DNA Database Composition

There are forensic databases in the majority of nations, however, the requirements for participation differ substantially. The types of offenses for which profiles will be included are often confined to the most serious ones, and profiles are only stored in the database for a short period of time. In most European nations, for example, the age of criminal culpability is between 13 and 16, hence younger children are not included. The most comprehensive database available is in the UK as given in Table 9.1. Most countries have their own legislation and regulations that define which samples to keep in the National DNA database. But most of them have the convicted and suspects stored in the database as a compulsory step. Other categories are not mandatory to store. Some countries have an elimination database for crime scene investigators and the witnesses. It is up to each country to decide whether offenses call for DNA profile submission to databases and whether or not to include arrestees. By using a DNA database, they have used the principle of proportionality to overrule people's right to autonomy in order to increase public safety (Wickenheiser 2022).

9.4 National DNA Databases

The UK's National DNA Database (NDNAD), created in 1995, is both one of the longest established, and biggest of such forensic DNA databases internationally. As of December 31, 2021, the UK National DNA Database had 6,829,278 subject profiles and 678,367 crime scene sample profiles (Amankwaa and McCartney 2019). While the UK hit rate has been cited to be above 60%, specific statistics are not available. The New Zealand DNA Profile Databank possesses over 200,000 samples and over 40,000 DNA profiles from forensic samples. There is a match rate of nearly 70% for all previously unsolved cases successfully linked to individuals and 30% link to another crime. As a result, its success can serve as an example of the potential of databases for solving crimes and any related issues. Naturally, this percentage will take into account encounters with victims and other individuals who were at the crime site for lawful

Table 9.1 NDNAD-England and Wales: policy on DNA profiling, storage, and use (Levitt 2007)

	Convicted offenders	Suspects	Arrested but not charged or charged but not acquitted	Young people	Volunteers (elimination samples)
Serious crimes	✓	✓	✓	✓	–
Minor offenses	✓	✓	✓	✓	–
Consent?	×	×	×	×	✓
Full samples retained	✓	✓	✓	✓	✓
Retained until death/age 100	✓	✓	✓	✓	✓
Speculative searching	✓	✓	✓	✓	✓
Familial matching	✓	✓	✓	✓	✓
Research	✓	✓	✓	✓	✓

purposes. When DNA crime scene samples are available, the detection rate for crimes rises. For instance, residential burglary went from a nationwide detection rate of 16% to 41% and car theft from 8% to 63% (Levitt 2007). Despite an increase in the number of individual profiles being stored, only roughly 0.36% of crimes are discovered using DNA analysis. By doing DNA analysis at the crime scene and promptly identifying persons, the Forensic Science Service's deployment of a mobile laboratory should speed up the rate at which crime scene profiles are placed onto the NDNAD. The UK NDNAD may have directly resulted in convictions in high-profile crimes, but its wider influence on public security objectives is still difficult to measure (Amankwaa and McCartney 2019). As of October 2021, the U.S. National DNA Index (NDIS) contained over 14,836,490 offender profiles, 4,513,955 arrestee profiles and 1,144,255 forensic profiles (Wickenheiser 2022). Those forensic profiles recovered from biological materials deposited at crime scenes have produced 587,773 hits assisting more than 574,343 investigations, which roughly equates to a 51.37% hit rate. Under the thinking that larger databases solves more crimes, this implies that roughly 556,482 forensic profiles, or 48.63% of them, are still pending for a match to provide a clue that will help police solve the murder. Instances in which there has not been a database match present a chance to think about other steps to give an investigative lead (Wickenheiser 2022). A larger database will produce more hits, but adding more profiles from the population of actively committing criminals to the database is crucial to improving database efficacy. Walsh et al. showed that while there are some apparent trends, there are no meaningful connections between the hit rate or return index (the number of hits divided by the number of samples) (Walsh et al. 2008).

China's database apparently contains more than 80,000,000 profiles of boys and men. The reality that men commit more crimes than women inspires this widespread collection. However, a particularly horrific crime wave in the Inner Mongolia region of Northern China, which resulted in the rape and homicide of 11 women and children, was what really spurred database expansion. Almost 230,000 fingerprints and over 100,000 DNA samples were gathered and examined in these cases (Wee 2020). These cases and the success of the investigation including DNA spurred state media to push for the creation of a national DNA database (Wickenheiser 2022).

9.5 Solving Cold Cases

The reopening of "cold" cases has been made possible by the capacity to create a DNA profile using lower quality and fewer amounts of DNA, even a single hair. A very positive return on investment using a single cold case project in Louisiana has demonstrated the direct expansion of DNA databases by adding offender profiles. Three American cold case programs, including Project Resolution from Louisiana, projects in Detroit, Michigan and Palm Beach County, Florida, have hit rates for forensic samples that range from 36% to 57.96% (Crouse et al. 2019; Wickenheiser 2021). This raises the issue of what can be done to help with the 42–64% of instances that are still unsolved. Resolving cold cases is an important contributor to preventing crime, increasing public safety and promoting trust (Barcus et al. 2019)

It has been established that a number of sorts of evidence that courts have traditionally depended on are less trustworthy and can result in erroneous convictions. Based on the first 325 DNA exonerations, The Innocence Project calculates that 72% of erroneous convictions featured misidentifications of eyewitnesses, 47% involved faulty or invalid forensic work, 27% involved police coercion of false confessions, and 15% involved shaky informant testimony (Dowdeswell 2022). Several innocent people had their lives seriously harmed and cast in doubt over the decades that the crimes in this sample remained unsolved. A factually innocent person who is found guilty of a crime committed by another offender is said to have been wrongfully convicted (Cassell 2018). Even apart from the tremendous injustice of a period of jail, wrongful convictions can result in severe stigmatization, anguish, and financial difficulty. Four persons were falsely convicted for the offenses in this sample; one, Henry Lee Lucas, received a death sentence that was later commuted. For suspects, victims, and their families, wrongful convictions can be traumatizing and difficult; when a conviction is reversed, victims and survivors must deal with the lack of closure and the knowledge that the real offender is still at large.

The rate of erroneous convictions has been the subject of various studies in the literature. Most scholars have used the number of exonerations as a proxy for the number of unjust convictions because the actual guilt or innocence of any individual is precisely what we do not know and precisely what the criminal process is designed to resolve. From peaking in the 1980s and early 1990s and peaking in 1995, exonerations have dropped (Gross et al. 2014). Because there is so much pressure on police and prosecutors to obtain a conviction, cases involving rape and murder may have greater rates of erroneous conviction than violent offenses as a whole (Cassell 2018). When Risinger examined capital rape-murder cases in the United States in the 1980s, he discovered a rate of wrongful conviction for the factually innocent of approximately 3.3%, with a maximum of almost 5% (Risinger 2006). These examples represent the sexually motivated homicides in the sample, which occurred in an era when such crimes were more prevalent than they are now. Risinger's rate of erroneous convictions is strikingly comparable to the 3.5% percentage of erroneous convictions for rape-murders discovered in the current dataset (Risinger 2006). So far, the FGG evidence has helped two people who were unfairly convicted get their sentences overturned. This collection contains several cases that have been open for very long times. The sample's rates of erroneous conviction varied from 2.1% of all offenses to 3.8% of all rape-murders, which are comparable to percentages seen in similar situations. The percentage of all cases that were wrongfully prosecuted ranged from 10.1% of all offenses to 16.5% of all rape-murders (Dowdeswell 2022)

9.6 DNA Mixture Profiles

Recent years have seen the expansion of the number of loci included in STR typing kits andstandardization of core loci across jurisdictions, allowing for greater cross-border sharing of DNA profiling data (Amankwaa 2020). However, it is a must to confirm the DNA match by retesting the reference samples with the same set of STRs in order to confirm the DNA hit as shown in Fig. 9.1.

Most experts agree that the likelihood ratio (LR) is the most potent and pertinent statistic for determining the weight of the DNA evidence (Gill et al. 2006). It is the ratio of the likelihood that the crime stain observed (O) will occur given

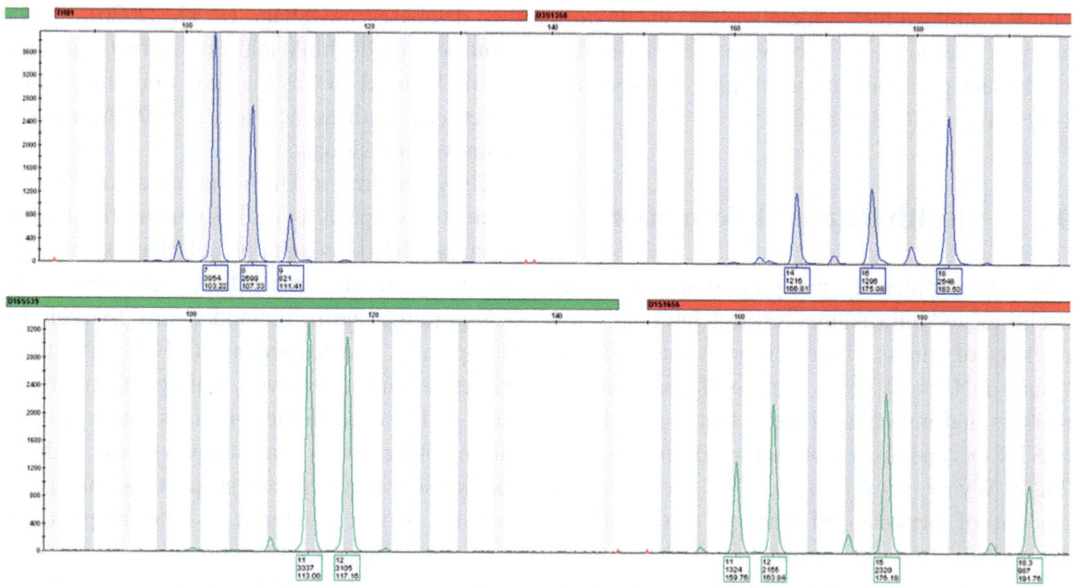

Fig. 9.1 Electropherogram showing part of a mixed DNA profile at four STR loci. Across a whole profile, interpretation of the varying number of peaks and peak heights can become very complex, even with a small number of contributors. (Source: Daeid et al. 2021)

each of the two competing hypotheses (H1 and H2) and all of the available information (I). Mathematically, we can put it like this:

$$LR = Pr(O|H1, I)/Pr(O|H2, I)$$

Normally, database search algorithms only count the number of concordant and discordant alleles rather than calculating an LR. Nonetheless, LR-based methods have been successful in locating relatives. Interpretation can be made more difficult by stochastic occurrences like heterozygote imbalance, allelic dropout, locus dropout, and allelic drop. When stutter peaks in mixed DNA profiles are similar in height to the minor allelic peaks, the ambiguity in the numbers of contributors and stutter, a by-product of the PCR process, might further make it difficult to evaluate the profile. A mixed DNA profile's contributor count can never be determined with precision. If the whole number of alleles is known, the answer could be found quickly. The uncertainty comes from the process of estimating the number of alleles from the peaks (Bright et al. 2013).

When two or more alleles from the same or different people are superimposed, it results in certain peaks that are not allelic (such as artifacts or stutter peaks). Due to dropout, certain alleles might not have formed a peak. It is probable that certain peaks with high sensitivity are caused by drop-in alleles, or alleles from the lab setting.

Because of the intricacy of the statistical calculations entailed in mixture interpretation methods, specialized software is necessary to do these analyses, and a variety of programs are being offered for this purpose (summarized in (Butler and Willis 2020)). As a result of these programs' ability to analyze mixtures that were once thought to be too complex for interpretation, forensic laboratories have adopted them quickly. Studies reporting on the creation and internal validation of various probabilistic genotyping software packages have also been published, along with recommendations for their use from a number of regulating bodies. Developers have urged forensic laboratories to make sure their analysts have a solid understanding of the concepts underlying the methods, that they continue to be involved in the interpretation of profiles, and that they critically assess the mixture analysis because the software packages that implement probabilistic genotyping methods are very complex (Gill et al. 2015)

9.7 Health-Related Information

Also, the profiles do not purport to now offer health-related information, research has uncovered markers of illness vulnerability in the so-called "junk" DNA and genes expressed near forensic markers and investigated the potential association of STR loci and disease phenotype (Bañuelos et al. 2022; Wyner et al. 2020). This increases the amount of personal information that could be extrapolated from the sample and would be useful to researchers, even though it would not aid in crime-solving.

9.8 Indirect Matching

Biologically related people naturally share DNA, therefore utilizing indirect matching techniques like partial matching, familial searching, and investigative genetic genealogy (IGG) also known as forensic genetic genealogy (FGG) which involves generation of a single nucleotide polymorphism (SNP) profile from an unsolved forensic case sample and entering it into a consumer genealogy database, the theoretical size of a DNA database can be enhanced. Searching for close matches that might suggest that someone on the database is genetically related to the crime suspect broadens the application of forensic databases even more. Such family history research could make a lot of individuals suspicious. In the DNA Forensic Lab in the Kingdom of Bahrain, they have come up with a technique that is called homozygous loci search (Al-Snan et al. 2020). Due to the fact there are a lot of consanguineous marriages in the Arab and Muslim regions in the world, results include in an increased number of homozygous runs. Thus, some families share the same homozygous loci in their profiles for generation making it easy to identify the unknown crime scene samples from a closely related individual DNA profile entered in the database (Al-Snan et al. 2020). Some countries might think that only direct matching should be used, just as they can think that different criteria justifies inclusion in a DNA database of known individuals.

Using the trait of DNA sharing among family to identify suspects as well as unidentified human remains (UHRs) is an approach that expands the scientific reach of DNA databases by incorporating indirect comparisons to seek for biologically related persons (UHRs). As these crimes remain unresolved, failing to investigate them could result in the criminals' at-large committing other crimes against new victims. A direct comparison between the suspect sample taken from the person of interest and the sample taken from the crime scene follows every indirect matching approach and acts as the final confirmation or rejection of the investigation lead.

9.9 Cases Solved by Indirect Matching

- American John Bittrolff, who beat two women to death, is a suspect in the Long Island Serial Killer Case and a convicted murderer. Rita Tangredi was discovered dead on November 2, 1993, while Colleen McNamee was murdered on January 30, 1994, after being battered and strangled. Partial Matching was used in these two incidents to identify John Bittrolff as a suspect. Timothy Bittrolff, his brother, was listed on the Combined DNA Index System (CODIS) for breaking a protective order. After that, a forensic sample discovered in 2013 on the remains of Rita Tangredi and Colleen McNamee was indirectly matched to Tim Bittrolff's profile. Kinship analyses confirmed that a biological relationship between the comparable non-matching profiles might have some merit. A report with the investigative lead was released following this familial assessment. Investigators discovered Timothy Bittrolff had a sibling who suited the circumstances of the two crimes, including his close proximity to the crime scenes, after they were notified that he was probably the killer's close relative. They got a DNA sample from John, Tim's brother. John's DNA was examined and precisely matched to forensic profiles discovered on both homicide victims. After being found guilty, John Bittrolff was sentenced to two consecutive terms of 25 years in prison (Wickenheiser 2022).

- Another case that was solved by indirect matching is a terrorism case in Bahrain, where a juvenile used Molotov bottle to throw at a police check point causing some casualties to a police man. The tape found on the Molotov bottle generated unknown male DNA profile. The seven homozygous loci were entered in the DNA database which gave partial hit with a known person, who was found it to be the sibling of the wanted juvenile (Al-Snan et al. 2020).
- More than 20 years after a 13-year-old girl body was discovered in a garbage in the Bronx, New York, authorities made an arrest on murder charges on November 29, 2021. The search for Minerliz Soriano began on February 24. Four days later, her body was discovered in a dumpster, wrapped in plastic. Semen found on her body was used to create a forensic profile that was entered into CODIS but did not produce a database hit. Familial searching was requested by the investigators in an effort to crack the case. A first order relative received a lead in their investigation from familial searching. With that relative, the police created a tree of their immediate relatives, which contained the murder suspect. When a sample of the suspect was taken, it exactly matched the forensic profile of the semen taken from Minerliz Soriano. There will be a trial (Wickenheiser 2022).
- The Golden State Killer, who was later identified as Joseph DeAngelo, was one of the first cases using IGG technique. Between 1974 and 1986, DeAngelo was responsible for over 13 deaths, over 50 sexual assaults, and over 200 burglaries. Significantly, the case was solved after 68 days of genealogical investigation that strengthened the work of the detectives. Throughout more than 40 years of conventional investigation prior to the application of the improved capabilities of DNA indirect comparison, DeAngelo had managed to elude detection. The ability of investigators to include and rule out suspects by casting a wider net than direct comparison to known samples is greatly enhanced by the increased use of DNA tools, particularly indirect matching techniques (Wickenheiser 2022). While the 20 STR core CODIS loci's low discriminating power typically restricts relationships to close relatives, IGG can link fourth or fifth cousins to the forensic profile.

The ability of investigators to include and rule out suspects by casting a wider net than direct comparison to known samples is greatly enhanced by the increased use of DNA tools, particularly indirect matching techniques. The proposed novel combination and expansion of existing methodologies has the potential to uncover new, unrecognized familial connections, which would then produce new investigation leads to resolve unresolved cases. The idea is to employ DNA regions that can be compared to one another, including Y-STRs, mtDNA, and X Chromosome profiles, for matching. Since they will have the same Y-STR and X Chromosome profiles (for males) if related paternally or the same mtDNA if related maternally, this comparison will favor related individuals.

These regions of common DNA are used as a search key to enable a second comparison of people who share the same key, allowing a kinship analysis to determine how closely related they are. Indirect matching, kinship analysis, and direct Y-STR and mtDNA comparison techniques are frequently used in forensic situations. Genealogical databases enable searching based on X and Y chromosome information (Kaye 2013). This novel mixed flow represents a new use of these established approaches (Ge and Budowle 2021).

9.10 Conclusion

The goal of the forensic laboratory is to increase the value of the evidence, which is accomplished by creating DNA profiles that can be connected to known individuals (Wickenheiser 2021). The best evidence must be provided by forensic science service providers, who must maximize the specificity, possibility of producing a favorable comparison, and timeliness of the evidence gathered at crime scenes. By employing shared DNA as a search tool, indirect matching methods have successfully increased the DNA database's scientific scope. When a direct comparison between a person's known DNA profile and the forensic profile is made, it is possible to triangulate to the person who deposited the DNA by using related individuals as additional leads. The value of forensic evidence rises as a result of the use of indirect matching, which enhances the possibility of providing an association. Also, reopening cold cases have helped solved mysteries of unsolved cases or wrongfully convicted individuals.

References

Al-Snan NR, AlBuarki FJ, Sayed SS (2020) The effect of consanguineous marriages in solving DNA cases. In: Dash HR, Shrivastava P, Lorente JA (eds) Handbook of DNA profiling. Springer, Singapore, pp 1–13

Amankwaa AO (2020) Trends in forensic DNA database: transnational exchange of DNA data. Forensic Sci Res 5:8–14

Amankwaa AO, McCartney C (2019) The effectiveness of the UK national DNA database. Forensic Sci Int Synerg 1:45–55

Bañuelos MM, Zavaleta Alennie YJA, Roldan Reyes RJ, Guardado M, Chavez Rojas B, Nyein T, Rodriguez Vega A, Santos M, Huerta-Sanchez E, Rohlfs RV (2022) Associations between forensic loci and expression levels of neighboring genes may compromise medical privacy. Proc Natl Acad Sci 119(40). https://doi.org/10.1073/pnas.2121024119

Barcus H, Heurich C, Ritter N, Schwarting DE, Walton R (2019) National best practices for implementing and sustaining a cold case investigation unit. Annotation

Bright J-A, Taylor D, Curran JM, Buckleton JS (2013) Developing allelic and stutter peak height models for a continuous method of DNA interpretation. Forensic Sci Int Genet 7(2):296–304

Butler JM, Willis S (2020) Interpol review of forensic biology and forensic DNA typing 2016–2019. Forensic Sci Int Synerg 2:352–367

Cassell PG (2018) Overstating America's wrongful conviction rate: reassessing the conventional wisdom about the prevalence of wrongful convictions. Ariz L Rev 60:815

Crouse CA, Bauer L, Sessa T, Looper A, Sikorsky J, Yeatman DT (2019) Combined DNA index system (CODIS)-based analysis of untested sexual assault evidence in Palm Beach County Florida. Forensic Sci Int Synerg 1:253–270

Daeid NN, Hackman L, Haddrill PR (2021) Developments in forensic DNA analysis. Emerg Topics Life Sci 5(3): 381–393

Dowdeswell TL (2022) Forensic genetic genealogy: a profile of cases solved. Forensic Sci Int Genet 58: 102679

Ge J, Budowle B (2021) Forensic investigation approaches of searching relatives in DNA databases. J Forensic Sci 66(2):430–443

Gill P, Brenner CH, Buckleton JS, Carracedo A, Krawczak M, Mayr W, Morling N, Prinz M, Schneider PM, Weir B (2006) DNA commission of the International Society of Forensic Genetics: recommendations on the interpretation of mixtures. Forensic Sci Int 160(2–3):90–101

Gill P, Haned H, Bleka O, Hansson O, Dørum G, Egeland T (2015) Genotyping and interpretation of STR-DNA: low-template, mixtures and database matches—twenty years of research and development. Forensic Sci Int Genet 18:100–117

Gross SR, O'brien B, Hu C, Kennedy EH (2014) Rate of false conviction of criminal defendants who are sentenced to death. Proc Natl Acad Sci 111(20): 7230–7235

Guillén M, Lareu MV, Pestoni C, Salas A, Carracedo A (2000) Ethical-legal problems of DNA databases in criminal investigation. J Med Ethics 26(4):266–271

Kaye DH (2013) The genealogy detectives: a constitutional analysis of familial searching. Am Crim L Rev 50:109

Levitt M (2007) Forensic databases: benefits and ethical and social costs. Br Med Bull 83(1):235–248

Ortner S, Whitehead H (1992) Introduction: accounting for sexual meanings. In: Sexual meanings: The cultural construction of gender and sexuality. Cambridge University Press, Cambridge, pp 1–27

Risinger DM (2006) Innocents convicted: an empirical justified factual wrongful conviction rate. J Crim L Criminol 97:761

Teodorović S, Mijović D, Nenadić UR, Savić M (2017) Attitudes regarding the national forensic DNA database: survey data from the general public, prison inmates and prosecutors' offices in the Republic of Serbia. Forensic Sci Int Genet 28:44–51

Thibedeau AD (2011) National forensic DNA databases. Council for Responsible Genetics, Cambridge, MA

Van Camp N, Dierickx K (2008) The retention of forensic DNA samples: a socio-ethical evaluation of current practices in the EU. J Med Ethics 34(8):606–610

Voultsos P, Njau S, Tairis N, Psaroulis D, Kovatsi L (2011) Launching the Greek forensic DNA database. The legal framework and arising ethical issues. Forensic Sci Int Genet 5(5):407–410

Walsh SJ, Buckleton JS, Ribaux O, Roux C, Raymond T (2008) Comparing the growth and effectiveness of forensic DNA databases. Forensic Sci Int Genet Suppl Ser 1(1):667–668

Wee S-L (2020) China is collecting DNA from tens of millions of men and boys, using US equipment. The New York Times 17

Wickenheiser RA (2021) Reimagining forensic science—the mission of the forensic laboratory. Forensic Sci Int Synerg 3:100153

Wickenheiser RA (2022) Expanding DNA database effectiveness. Forensic Sci Int Synerg 4:100226

Williams R, Johnson P (2005) Inclusiveness, effectiveness and intrusiveness: issues in the developing uses of DNA profiling in support of criminal investigations. J Law Med Ethics 33(3):545–558

Wyner N, Barash M, McNevin D (2020) Forensic autosomal short tandem repeats and their potential association with phenotype. Front Genet. https://doi.org/10.3389/fgene.2020.00884

10 Guidelines, Ethical Issues, and Other Challenges of Forensic DNA Analysis

10.1 Introduction

The advantages of DNA profiling are obvious. It suggests that the DNA samples came from different people when two DNA profiles do not match; this is known as exclusionary evidence (Gans and Urbas 2002). Post-conviction many people who were wrongfully convicted have been freed due to DNA comparison (Walsh 2009). The gold standard of forensic scientific evidence is now forensic DNA analysis, which is based on the examination of a single or straightforward mixture of DNA samples. DNA profiles from crime scenes can be matched to national DNA databases, which are present in the majority of modern nations. It is now possible to analyze complex DNA mixes, including low copy number (LCN) samples and DNA from three or more persons. It is now possible to go beyond the DNA's "non-coding" regions because of recent genetics research and technological advancements. By connecting a suspect's DNA profile to a biological family member whose profile is in the database, familial searching can now be used to locate potential suspects whose DNA profiles are not in a particular national database. Additionally, using DNA phenotyping, one may forecast a person's outward appearance, including their eye, hair, and skin color (as well as other genetic traits), and deduce their biogeographical ancestry. These new developments have reignited discussions about possible violations of human rights. Politicians and law enforcement organizations have viewed DNA evidence as essential to criminal investigations since DNA analysis was first included into the criminal justice system. This point of view has influenced the growing use of DNA analysis, which, together with other branches of forensic science, is crucial to the criminal justice system. However, Corns asserts that there have been issues in the link between science and legislation (Corns 1990).

10.2 Debate of Forensic DNA Analysis in the Past

In England's criminal court system, DNA profiling was first used in 1986. The body of a 15-year-old schoolgirl was discovered in 1983 in the English village of Narborough, close to Leicester. Before being killed, she had suffered a sexual assault. The death of a second 15-year-old girl was not discovered for 3 years, during which time the case remained unsolved. As a consequence of DNA "fingerprinting," Pitchfork was captured and was the first person to be found guilty. Following the Pitchfork case, other jurisdictions used DNA analysis, which sparked debate and controversy. Questions about its fitness and acceptability became the basis of several academic and legal discussions. In the analysis of the Australian literature on the subject of the relationship between DNA technology and the law, Corns noted the emergence of three "models of interpretation" (Corns 1990: 25). These

H. R. Dash et al., *Advancements in Forensic DNA Analysis*,
https://doi.org/10.1007/978-981-99-6195-5_10

arguments have been presented in several probes, academic and scientific papers, and admission hearings (Lynch 2013). They were mostly focused on disagreement on the scientific veracity of the findings. Arguments included a focus on the creation of an all-encompassing and trustworthy technical method for analyzing and presenting scientific evidence in court. Certain incidents exemplify the key problems of the legal, scientific, and libertarian frameworks (Corns 1990).

10.3 The Legal Model

The legal model concentrated on the implications of new forensic DNA analysis technology within the preexisting systems of criminal investigation and trial. This model's major concern was the conviction of the guilty. It suggested that any increase of police powers to accommodate the new DNA technology was required. According to Corns (1990), the legal model arose out of a necessity for criminal control. The requirement to prosecute the guilty was itself a rationale for the use of DNA analysis and the extension of police powers related to that use. This model emphasized the advantages that DNA analysis would have for the criminal justice system, particularly for the prosecution of criminals, and it urged for laws to make it easier to use DNA evidence in criminal cases. As one of two competing criminal justice models—the other being the due process model—Packer introduced the crime control model. The crime control model is based on a value system that contends that the suppression of crime is the primary duty of any criminal justice system. The model calls for an efficient system that enables suspects to be screened, guilt to be established, and convictions to be obtained. According to Packer, the ability to identify innocence or guilt early on "by the application of administrative expertness, primarily that of the police and prosecutors" is crucial for the successful conclusion of the criminal justice process (Packer 1964: 11). Through informal administrative fact-finding by the police and prosecution, an early "presumption of guilt" can be formed, which acts as a prediction of the outcome of the case. In order to accomplish this, "it becomes important to place as few restrictions on the character of the administrative fact-finding process as possible and to limit restrictions to those that enhance reliability" (Packer 1964: 13).

According to Corns (1990), DNA technology has become a significant political tool for reducing crime. He cites the Crimes (Blood Samples) Act 1989 (Vic) as an illustration of the political movement in Victoria, Australia, toward crime control. According to suggestions made by the Victorian Consultative Committee on Police Powers of Investigation, this legislation was created to give Victorian police officers the ability to draw blood from suspects in crimes in order to compare it to DNA from the crime scene. According to Corns, the advancement of DNA profiling technology served as the impetus for passing this law, which breezed through parliament with bipartisan support.

10.4 The Scientific Model

While the scientific model examined DNA technologies in terms of the accompanying testing processes, probability theories, sample errors, and contamination potential (Corns 1990), the former approach was more realistic. In other words, the model expected adherence to the scientific model and the scientific validation of the operations because it was concerned with science and scientific terminology. The use of DNA as evidence in court in the United States was met with considerable controversy as molecular geneticists argued with the forensic scientists and attorneys who were using this kind of evidence. Laboratory scientists who were largely untrained in the procedure employed the fundamental processing method, restriction fragment length polymorphism, or RFLPs, created by Jeffreys, more frequently in the criminal court system without any regulation or control. The use of probability and non-standardized interpretation was also questioned. Estimating the likelihood of a random match within a population presented challenges (Lander 1991) particularly for sub-populations. This raised concerns about whether probability computations may be used as evidence in court (Edmond 2011). Human geneticists "have a dual responsibility to society: to develop new tools and to ensure that they are used to the highest standard," as noted by Lander (1991: 822). The basic RFLP analysis took a lot of time and money to do; it required a double strand of DNA, therefore in order to create a DNA profile, a sizable, high-quality DNA sample was necessary, which is hard to find at crime scenes; also, the results were sometimes ambiguous. The polymerase chain reaction, or PCR, a technique that can amplify DNA in a single cell, was one of the first scientific innovations to be used in forensic DNA analysis. This was followed by the examination of short tandem repeats, or STRs (Jeffreys 2005).

10.5 The Libertarian Model

The libertarian model assessed the potential advantages of DNA profiling against the effects it might have on established legal norms and accused persons' rights. It cast doubt on the veracity of the scientific approach (Hocking et al. 1997). Although a positive DNA match may suggest that two samples came from the same person, Gans and Urbas emphasized in a 2002 article that there are other possibilities for a match (Gans and Urbas 2002). First, coincidental matches, where the sample originates from an unrelated person with the same profile, are possible (noting that this issue is more common with the use of fewer loci). Second, kinship relationships can result in matches, which means that the DNA originates from a related person. Third, a match can be the result of contamination—that is, the crime scene sample may have picked up DNA from someone unrelated to the case. According to Thompson, Taroni, and Aitken (Thompson et al. 2003), a false positive can offer a different explanation for a match just like a coincidental match does. A false positive may originate from a mistake in the sample's handling, in the sample's collection, in the interpretation of the test results, or in the reporting of the results. Alternative explanations for DNA matches raise the danger of erroneous convictions brought on by a reliance on DNA evidence that is too great. From a human rights perspective, this possibility makes the careless use of DNA analysis less desirable. Invasion of human rights and ethical considerations can result from the use of DNA analysis in the criminal justice system in a number of different ways. This viewpoint is consistent with Packer's due process model (Packer 1964), which is the second paradigm of criminal justice. The due process model is skeptical of an administrative and investigative process without judicial scrutiny since it relies on the presumption of legal innocence. It disapproves of casual fact-finding and guilt assessments in favor of a formal adversarial determination of guilt (Packer 1964).

10.6 Forensic DNA Analysis

In an effort to settle the controversy surrounding the use of DNA evidence in court, the United States (US) National Research Council (NRC) looked into the reliability, validity, and secrecy concerns that emerged regarding the use of DNA analysis in 1992. The investigation's goal was to improve forensic DNA technologies for use in the criminal justice system. The report's objective was to describe the fundamental science and technology behind DNA typing (Council 1992). The study took into account the methods employed at the time for DNA typing, including PCR and restriction enzymes. There have been concerns expressed about the validity of the technical methods employed, including the lack of quality control among various laboratories and the match criteria. The accuracy of the DNA analysis as it was presented in court, including the use of population statistics to calculate a random inclusion, was contested. It was noted that DNA evidence, in contrast to fingerprint evidence, includes more personal information, and as a result, questions about confidentiality and privacy were raised regarding its use in the criminal justice system. Despite these reservations regarding the use of DNA typing, the report suggested that DNA typing should continue to be developed and used for forensic purposes and that the existing laboratory procedures for identifying DNA variations were essentially sound (Council 1992).

The results of the 1992 study were examined in a second report on forensic DNA technology that the NRC produced in 1996. It confirmed that the DNA technology was reliable and that, provided quality-controlled procedures were used, the findings were very reproducible. According to the report, STRs are the most promising newer approach to be used in DNA typing, and STRs are expected to eventually replace the RFLP method (Council 1996). The most reliable form of forensic science proof eventually emerged: forensic DNA analysis. The need for a management structure to gather, store, and share the various DNA profiles became a practical fact with the increased use of DNA evidence in criminal cases, and the majority of developed countries introduced DNA databases through the enactment of the required legislation (Gabel 2010).

A thorough DNA record was first released in 1995 by the United Kingdom, then in 1996 by New Zealand. As a pilot project, the United States developed the Combined DNA Index System (CODIS) in 1990 to bring together the different national, state, and local DNA databases. Although US law authorized the development of a national DNA index system in 1994 (Gabel 2010), the CODIS (NDIS) national level was not operational until 1998. Following the passage of Part 1D of the Crimes Act 1914, CrimTrac created the National Criminal Investigation DNA Database (NCIDD) in Australia in 2001 (Cth). All Australian states' sharing of DNA profiles was made possible by the NCIDD (Smith 2019).

A worldwide survey offers a thorough database of data on the status of DNA databases and legislation for more than 100 nations (Wallace et al. 2014). Sixty-nine of the member nations of Interpol reported having a national DNA database in 2017 according to a global DNA profiling study (Butler and Willis 2020). Eighty-nine of the 110 National Central Bureaus that replied to Interpol's survey of 194 Bureaus in 2018 reported using DNA profiles in police investigations, and 70 nations had a searchable DNA database. This included seven African nations, 10 American nations, 13 Asian nations, and 40 European nations (Oosthuizen and Howes 2022).

Several inherent restrictions are present despite the elevated standing of DNA and advancements in its storage and utilization. For instance, a DNA match cannot prove an accused person is guilty (Corns 1992); it can only corroborate that a person may have been present at the scene of a crime (Hocking et al. 1997). Additionally, trace DNA and the interpretation of mixed samples present difficulties. DNA is also prone to environmental degradation. DNA that is typically found on items at a crime scene but cannot be linked to a specific body fluid is referred to as trace or touch DNA. This DNA may be limited in quantity or quality or was discovered using the low copy number (LCN) DNA method, depending on the circumstances (Meakin and Jamieson 2013). Direct trace DNA can be left without any interaction, for example, by depositing DNA while speaking or coughing (Port et al. 2006). Trace DNA can also be transferred indirectly or secondarily and left at a crime site. The DNA of the other person may be transferred onto an object at a crime scene, for instance, even though that person was not present at the scene when two people shake hands and transmit their DNA to one another or an object thereafter (Lowe et al. 2002; Meakin et al. 2017).

DNA samples that may have been damaged by environmental factors are referred to as degraded samples. A DNA sample may be degraded by being broken down into smaller pieces by a variety of environmental variables, including water, oxygen, sunlight exposure, and various environmental enzymes. Degraded samples were extremely challenging, if not impossible, to analyze using RFLPs, but multiplex STR and low copy number (LCN) DNA methods work better for their detection (Butler 2011).

LCN of DNA, such as cells from a fingerprint, which may contain as few as five cells, is a sophisticated biological sample from which DNA can be extracted (Wise 2009). After the DNA has been extracted from the cells, it is replicated using PCR. To guarantee a sample large enough for STR sequencing, the PCR cycles must be increased. LCN DNA has a number of issues. The likelihood that part of the sequence will be lost during the increased number of amplifications of the tiny amount of DNA increases, potentially leading to an incorrect profile. A small sample size raises the possibility of contamination, either in the course of the inquiry or in the lab. Additionally, the DNA is lost in the PCR process because of the sample size, which means DNA is not reserved and the procedure cannot be replicated for reanalysis. LCN raises the frequency of mixed samples, where the biological material includes DNA from multiple people, making it more difficult to produce a distinct profile.

10.7 Evolution of DNA Techniques

Technological developments include DNA methylation pattern changes brought about by epigenetic adaptations of the genome and massively parallel sequencing (MPS), which uses both STRs and SNPs (Butler and Willis 2020). They enable the extraction of more precise and thorough information from the DNA sample, including lineage informative markers and an individual's outward physical characteristics (Hasnain 2019). SNPs can be used when degraded materials or low copy numbers make STR typing impossible. A single nucleotide polymorphism (SNP) is a difference between people in a single position of a DNA sequence. A recent study in technical capabilities made it possible to identify up to ten distinct contributors in a DNA mixture using massively parallel DNA sequencing, which used thousands of SNPs with low minor allele frequencies (Petrovick et al. 2020).

A DNA profile from a crime scene can be linked to a suspect using the DNA of a near relative who shows in a database using familial searching because DNA profiles can show genetic relationships (Smith and Mann 2015). A familial search may produce partial matches in instances where a crime scene DNA profile is thought to be that of the offender but does not match any profile on the DNA database, providing leads for the investigation (Maguire et al. 2014). Familial searching has been used in numerous jurisdictions with success but is not allowed in all jurisdictions. Biological similarity is used in familial searching to identify potential culprits in criminal cases. It implies that some population segments, specifically those who are already disproportionately represented in the criminal justice system, will be at a greater risk of being investigated. Known racial disparities within the criminal justice systems will result in racially biased genetic population surveillance, according to Piquado et al., where a portion of the population is disproportionately represented within the criminal justices system (Piquado et al. 2019: 43)

Additionally, it is now feasible to determine external traits and features of a person's physical appearance from their DNA using the coding regions of the human genome (Kayser 2015). Without the aid of an eyewitness, DNA phenotyping enables the creation of an identikit of a suspect's potential exterior characteristics. In the US, authorities frequently release photofits of suspects based on forensic DNA phenotyping. Police in the US used genetic ancestry and DNA phenotyping in the well-known case of the Golden State Killer in their quest for justice (Wickenheiser 2019).

High-density SNP profiles and private datasets run by businesses like Ancestry.com are used in investigative genealogy (Ge and Budowle 2021). If a balance between privacy interests, proper oversight, and database composition is created, forensic genetic genealogy will offer various opportunities for Australian law enforcement to generate investigative leads, according to Scudder et al. (2020) Despite the fact that Australia does not frequently use DNA phenotyping, the Queensland Police Service reported using phenotyping services offered by a US-based firm in 2016 (Smith 2019).

Rapid DNA technology has emerged in response to the need for quicker DNA findings. In less than 2 h, a portable DNA device can create a DNA profile (Wilson-Wilde and Pitman 2017). The Rapid DNA devices are making their way to crime sites from remote locations like police stations and border crossings. Rapid DNA technology, on the other hand, uses the complete input sample and is less sensitive than conventional DNA analysis, which could lead to the loss of evidence. This means that if Rapid DNA is chosen to be used at the crime site, a decision may be made between a quicker analysis at the risk of losing evidence and a more thorough laboratory analysis at the risk of running out of time to catch the offender (Mapes et al. 2019)

10.8 Guidelines for DNA Analysis

To ensure the consistency and integrity of scientific research findings, many guidelines have been developed. However, no particular guidelines for the moral collection of such data have been developed. These guidelines provide details of the general processes for conducting ethical research on materials with human, animal, plant, and environmental origins as well as a summary of the generally accepted principles for conducting ethical research on biological materials.

Human biological material (HBM) refers to all biological material of human origin acquired for study by direct donation from an individual. This includes organs, tissues, bodily fluids, teeth, hair, and nails as well as DNA, RNA, and metabolites. The idea of an individual HBM was expanded to include a "interactive community of human cells and microbial cells" (Rees et al. 2018) because access to our microbiomes and pathogens is voluntary.

The manuscripts must state the following (D'Amato et al. 2020):

1. the ethics committee or regulatory board that authorized the research procedures and the design of the study before publishing any content with a human origin. All parties engaged must give their ethical consent in cases where the country(ies) of collection and analysis differ.
2. a certification that the protocols and approved processes were followed, and
3. a declaration that each participant signed the approved consent form, signifying their free and voluntary assent.

All participants' personal information must be safeguarded against sharing or public disclosure in any way. In accordance with the UNESCO declaration of human genome and human rights (Unesco 1997), the informed consent document contains clauses about the objectives and scope of the study, the risks and benefits of participating, disclosure of the fate and utilization of the individual's donated biological material, including the possibility that it will be shared with national and international collaborators, and to make appropriate provision for withdrawal from the study (D'Amato et al. 2020). Many institutions has institutional and human review boards that review proposals to collect and analyze human biological material as well as survey or educational data. The goal and impact of the research or study must outweigh potential risks or harms to participants.

The use of biological samples for purposes other than the main one for which the study was approved is known as "secondary use of biological material." The research is not approved for publication unless the extended use of samples was covered by the ethics approval and informed consent. The use of samples not derived from free will participation (e.g., prisoners of war, detainees by law enforcement agencies, adults without the mental capacity to provide consent, embryos) and participation of illiterate people in the form of video recorded verbal consent will not be accepted unless covered by national legislation, which must be indicated along with the corresponding permit. Legal guardians must give their consent for minors, as defined by the laws of their respective nation of origin, to participate (s).

Population genetics of the investigative field getting data on participants' genetic identities is a requirement of genetic study. Genetic and genomic data must be released to the public in accordance with the consent documents signed by willing donors and the general ethical standards adopted by the journals. Minorities, indigenous peoples, and vulnerable population groups may be involved in this sort of work. These individuals frequently have a distinct cultural background than the researchers. Each community should be treated with respect, taking into account its culture, traditions, and religious views (D'Amato et al. 2020). When these studies are carried out by multinational teams, local researchers are required to participate because they have a greater understanding of the populations and communities being studied. Prior to the papers being evaluated, documentation from all ethics review boards (from the nations that participated in the sampling to the nations that carried out the experimental work) will be required. Additionally, population composition may shift over time or in one direction due to human movements, migrations, and displacements, such as immigration into urban areas or large-scale migrations out of conflict zones. The date of collection must be stated by the writers because of this.

Even fewer explicit rules exist regarding access to human remains' biological material. According to some laws, the consent of the next of kin may be required for research on postmortem cases. Different boards, professional organizations, and legislative bodies may have rules governing who has access to human remains through exhumation or excavations (Heathfield et al. 2017). For instance, law enforcement may need permission from an archaeological organization to enter mass graves with forensic anthropology teams. We will need supporting evidence.

10.9 Post Publication Policies and Conscientious Objection

Manuscript submissions must be accompanied by proof that the publication rules have been followed. Original submissions will be rejected if there has been a violation of local laws, foreign agreements, or any ethical standards. If violations are discovered after publication, they may be retracted or otherwise rectified in accordance with the ethics policies of the journal after an inquiry and due process. "While informed consent to research is important, the mere possibility that a participant or surrogate may be willing to consent to research does not, in and of itself, mean that the research is ethically acceptable," the WHO guidelines state (D'Amato et al. 2020). For instance, recent scientific research on wildlife with the necessary ethics research permits was not allowed publication because the editorial board of the journal found the research to be unethical (Costello et al. 2016)

References

Butler JM (2011) Advanced topics in forensic DNA typing: methodology. Academic Press, Oxford

Butler JM, Willis S (2020) Interpol review of forensic biology and forensic DNA typing 2016–2019. Forensic Sci Int Synerg 2:352–367

Corns C (1990) DNA is watching. Arena 92:24–28

Corns C (1992) The science of justice and the justice in science. Law Context 10:7

Costello M, Beard KH, Corlett RT, Cumming GS, Devictor V, Loyola R, Maas B, Miller-Rushing AJ, Pakeman R, Primack RB (2016) Field work ethics in biological research. Biol Conserv 203:268–271

D'Amato ME, Bodner M, Butler JM, Gusmão L, Linacre A, Parson W, Schneider PM, Vallone P, Carracedo A (2020) Ethical publication of research on genetics and genomics of biological material: guidelines and recommendations. Forensic Sci Int Genet 48:102299

Edmond G (2011) Review essay: the building blocks of forensic science and law: recent work on DNA profiling (and photo comparison). Soc Stud Sci 41(1): 127–152

Gabel JD (2010) Probable cause from probable bonds: a genetic tattle tale based on familial DNA. Hastings Women's LJ 21:3

Gans J, Urbas G (2002) DNA identification in the criminal justice system. Trends and issues in crime and criminology. Australian Institute of Criminology, Canberra

Ge J, Budowle B (2021) Forensic investigation approaches of searching relatives in DNA databases. J Forensic Sci 66(2):430–443

Hasnain SE (2019) Forensic epigenetic analysis: the path ahead. Med Princ Pract 28(4):301–308

Heathfield LJ, Maistry S, Martin LJ, Ramesar R, de Vries J (2017) Ethical considerations in forensic genetics research on tissue samples collected post-mortem in Cape Town, South Africa. BMC Med Ethics 18(1):1–8

Hocking BA, McCallum H, Smith A, Butler C (1997) DNA, human rights and the criminal justice system. Aust J Hum Rights 3:208–237

Jeffreys AJ (2005) Genetic fingerprinting. Nat Med 11(10):1035–1039

Kayser M (2015) Forensic DNA phenotyping: predicting human appearance from crime scene material for investigative purposes. Forensic Sci Int Genet 18:33–48

Lander ES (1991) Research on DNA typing catching up with courtroom application. Am J Hum Genet 48(5): 819

Lowe A, Murray C, Whitaker J, Tully G, Gill P (2002) The propensity of individuals to deposit DNA and secondary transfer of low level DNA from individuals to inert surfaces. Forensic Sci Int 129(1):25–34

Lynch M (2013) Science, truth, and forensic cultures: the exceptional legal status of DNA evidence. Stud Hist Philos Sci Biol Biomed Sci 44(1):60–70

Maguire CN, McCallum LA, Storey C, Whitaker JP (2014) Familial searching: a specialist forensic DNA profiling service utilising the National DNA Database® to identify unknown offenders via their relatives—the UK experience. Forensic Sci Int Genet 8(1):1–9

Mapes AA, Stoel RD, de Poot C, Vergeer P, Huyck M (2019) Decision support for using mobile rapid DNA analysis at the crime scene. Sci Justice 59(1):29–45

Meakin G, Jamieson A (2013) DNA transfer: review and implications for casework. Forensic Sci Int Genet 7(4): 434–443

Meakin GE, Butcher EV, van Oorschot RA, Morgan RM (2017) Trace DNA evidence dynamics: an investigation into the deposition and persistence of directly-and indirectly-transferred DNA on regularly-used knives. Forensic Sci Int Genet 29:38–47

National Research Council (US) Committee on DNA Technology in Forensic Science (1992) DNA technology in forensic science. National Academies Press, Washington, DC

National Research Council (US) Committee on DNA Technology in Forensic Science (1996) The evaluation of forensic DNA evidence. National Academies Press, Washington, DC

Oosthuizen T, Howes LM (2022) The development of forensic DNA analysis: new debates on the issue of fundamental human rights. Forensic Sci Int Genet 56: 102606

Packer H (1964) Two models of the criminal process. U Pa L Rev 113:1

Petrovick MS, Boettcher T, Fremont-Smith P, Peragallo C, Ricke DO, Watkins J, Schwoebel E (2020) Analysis of complex DNA mixtures using massively parallel sequencing of SNPs with low minor allele frequencies. Forensic Sci Int Genet 46:102234

Piquado T, Matthies CF, Strang L, Anderson JM (2019) Forensic familial and moderate stringency DNA searches: policies and practices in the United States, England, and Wales. RAND, Santa Monica, CA

Port NJ, Bowyer VL, Graham EA, Batuwangala MS, Rutty GN (2006) How long does it take a static speaking individual to contaminate the immediate environment? Forensic Sci Med Pathol 2:157–163

Rees T, Bosch T, Douglas AE (2018) How the microbiome challenges our concept of self. PLoS Biol 16(2):e2005358

Scudder N, Daniel R, Raymond J, Sears A (2020) Operationalising forensic genetic genealogy in an Australian context. Forensic Sci Int 316:110543

Smith M (2019) Three stages in the development of DNA evidence in Australia. Aust J Forensic Sci 51(6): 705–714

Smith M, Mann M (2015) Recent developments in DNA evidence. Trends and issues in crime and criminal justice, vol 506. Australian Institute of Criminology, Canberra

Thompson WC, Taroni F, Aitken CG (2003) How the probability of a false positive affects the value of DNA evidence. J Forensic Sci 48:1

UNESCO (1997) Universal declaration on the human genome and human rights. Article 6. The General Assembly of the United Nations endorsed the UNESCO Declaration in Dec. 1998. United Nations Press Release GA/9532

Wallace H, Jackson A, Gruber J, Thibedeau A (2014) Forensic DNA databases–ethical and legal standards: a global review. Egypt J Forensic Sci 4(3):57–63

Walsh SJ (2009) Evaluating the role and impact of forensic DNA profiling on key areas of the criminal justice system (Doctoral dissertation)

Wickenheiser RA (2019) Forensic genealogy, bioethics and the Golden State Killer case. Forensic Sci Int Synerg 1:114–125

Wilson-Wilde L, Pitman F (2017) Legislative and policy implications for the use of rapid DNA technology in the Australian context. Forensic Sci Policy Manage Int J 8(1–2):26–37

Wise J (2009) The new scientific eyewitness: the role of DNA profiling in shaping criminal justice. VDM Verlag Dr Müller, Saarbrücken

11 Application of Forensic DNA Technology in Analyzing Real-Time Casework Samples

11.1 Introduction

The French criminologist Edmond Locard developed what is now known as the Locard's exchange principle, or "every contact leaves a trace," in 1910 (Locard 1930). This fundamental principle, when applied to traces in forensic biology, purports that biological agents will leave minute amounts of DNA-containing material upon any contact with items or other living things. Given the capability of modern forensic DNA analysis, the presence of DNA-containing material from a specific person may be proven beyond a reasonable question, just as biological agents may be individually identified by simultaneous analysis of a sufficient number of DNA polymorphisms (Gosch and Courts 2019). This DNA-based individualization, however, does not provide information on the type of contact that caused the biological material to be transferred from the location where a sample was gathered, i.e., it does not permit contextualization of the trace material and, as a result, does not allow for the reconstruction of how it had been deposited (Gosch and Courts 2019).

The study of micro-traces like touch DNA or telogen hair samples presents numerous difficulties. Despite the fact that DNA typing techniques have significantly increased in sensitivity over the past few years, further development is still needed to recover the minuscule quantities of biological material contained in micro traces. For instance, it can be challenging to separate the DNA of the victim and the offender in criminal cases where the offender frequently only provides trace amounts of touch DNA. Single particle collection is labor- and time-intensive but produces admixed profiles in contrast to full sampling of the evidence (Dierig et al. 2020).

A novel sampling approach is suggested for tapings in forensic casework in order to increase DNA yield and success rate while at the same time avoiding mixture creation. This is in addition to optimizing identification and collecting bio particles during the process. Improvements to DNA extraction, a crucial step in micro trace analysis, entailed assessing the effectiveness and DNA recovery of various extraction techniques, particularly magnetic bead-based technology. Low template residues appeared to be best suited for direct lysis techniques. Even the presence of inhibitors is permitted by recently created commercial kits. The successful improvement of embraced tools successfully reduces the time and labor required for processing micro traces (Dierig et al. 2020).

Since useful genetic information can be obtained even from the smallest samples, such as DNA from touch or biological traces, advanced typing achievement has effectively aided in the

H. R. Dash et al., *Advancements in Forensic DNA Analysis*,
https://doi.org/10.1007/978-981-99-6195-5_11

resolution of many cold cases (van Oorschot et al. 2010; Lim et al. 2016).

11.2 Sources of DNA

The type of cells that make up a trace or stain have been shown to produce different DNA yields, with skin contact samples containing mostly sloughed, enucleated, and thus DNA-free keratinocytes while trace amounts of blood or saliva contain higher amounts of mostly intact cells and thus larger quantities of undegraded DNA (Sun and Reichenberger 2014). The DNA content of skin contact samples may contain significant amounts of extracellular, partially degraded DNA from apoptotic epithelial cells (Alessandrini et al. 2003), sebaceous glands (Zoppis et al. 2014), or sweat glands (Quinones and Daniel 2012), which would explain why those samples' DNA yield and integrity are lower than those from samples from other body fluids (Kita et al. 2008)

"Touch DNA" refers to biological material that is transmitted when two people or an individual and an object come into physical touch (Wickenheiser 2002; Daly et al. 2012). It has long been believed that DNA from handled samples primarily came from skin cells that had shed, specifically keratinocytes from the epidermis' top layer (Farash et al. 2015). Keratinocytes undergo full differentiation to become corneocytes during the keratinization process, which results in the loss of their nuclei (Kita et al. 2008). Realistically, DNA could originate from a variety of different places, including the following: conglomerates of nucleated epithelial cells and keratinocytes that had released (Oleiwi et al. 2015), through interaction with various body regions, cells from bodily fluids (such as saliva, tear fluid, urine, and nasal secretion) transmitted to the hands (Zoppis et al. 2014), and fragments of cell debris (such as "stripped" nuclei) or cell-free DNA (cfDNA), which can be endogenous (such as from perspiration or sweating) or exogenous (such as to the hands or even the person themselves) (Stanciu et al. 2015). Growing forensic interest centers on the question of how a particular DNA-containing sample got to the location where it was recovered. Several studies have demonstrated that trace amounts of DNA can transfer to objects in indirect, secondary, or even higher order transfer scenarios rather than just through direct skin contact (Lehmann et al. 2013). In fact, the number of samples from "swabbed surfaces" intended to recover invisible "touch" DNA is on the rise in most forensic routine laboratories, and expert witnesses in court are increasingly challenged to judge, for example, whether an incriminating DNA sample matching a suspect could have been transferred to the crime scene inadvertently and to speculate on the likelihood of such an occurrence.

Micro trace analysis has a lot of difficulties because there are only very small amounts of DNA accessible. Amplification artifacts frequently cause stochastic effects like imbalanced peak heights, allele drop-out, and allele drop-in, making it more difficult to acquire complete profiles. The slightest contamination is also amplified by enhanced amplification techniques, making it challenging to interpret micro traces.

11.3 Collection Methods

Cutting, swabbing, or taping of evidence-related objects are typical touch DNA collection techniques. Fabrics are the best candidates for cutting (Petricevic et al. 2006), without losing any cellular traces. However, the advent of inhibitors, such as laundry detergent, is likely to happen. Additionally, original bits of evidence may frequently not be destroyed. The swabbing method involves applying moderate pressure while rubbing and rotating a cotton swab that has been previously moistened. A second dry swab would then be used to gather the remaining sample (Barash et al. 2010). The majority of the DNA is recovered when both samples are co-extracted (Pang and Cheung 2007). Compared to porous or adsorbing substrates, smooth surfaces are where swabbing is most helpful. In fact, cloth and porous or rough surfaces were the best candidates for taping. Nowadays, a variety of designs are offered, including dissolvable or insoluble coatings, sticky tape, adhesive film, and ready-to-use adhesive stamps (Barash et al. 2010). The target area could be covered with adhesive tape or film, which could then be raised once and recovered with a cover sheet. When applied directly at the crime scene before bagging, this application of adhesive film secures all traces at their original locations, which is especially important in felony cases where minor (offender) DNA traces are frequently only accessible. Thus, during transit to a laboratory, displacement or loss is hereby avoided (Goray et al. 2012) and the initial evidence may be examined. When untangled lint and fingerprints are gathered simultaneously, it is possible to apply multidisciplinary analysis to a single sample, such as comparative fiber or dactyloscopic examination. After that, the adhesive surface is either scraped off after the application of a solvent (such as ethanol) or sections of the film are cut out for DNA analysis (Barash et al. 2010). Adhesive stamps are made to completely eliminate the adherent so that DNA can be extracted.

11.4 DNA Extraction

The co-collection and processing of DNA from all, sometimes multiple, donors is a drawback of the earlier techniques. Despite the generally good chances for DNA typing in these samples, DNA separation is hardly ever fully feasible. For instance, the victim's DNA frequently outweighs the offender's biological material in instances of physical assault. Typically, the DNA profile of the main contributor (the victim) is overrepresented and may suppress the alleles of the lesser contributor (the offender). Laser microdissection could be used to separate male and female cells; flow cytometry or the DEPArray enable segregation based on morphological or epitopical traits. In an ideal world, the admixture of samples would be prevented before the worldwide collection of all biological material.

Another crucial step in forensic casework is DNA extraction, particularly when dealing with low template samples. Common extraction techniques include purification stages that reduce available DNA yields, sometimes to levels below detection. The magnetic beads extraction principle provides the chance to preserve the most important biological material possible for later analysis. Due to sample retention requirements and the push for dual STR typing in most forensic routine institutions, direct PCR as a method for enhancing findings in micro trace analysis was purposefully not considered.

11.5 Application of DNA Analysis

In this chapter, different type of approaches were discussed in cases such as terrorism, sexual assault, drugs, and burglary.

11.6 Anti-Terrorism Cases

Globally, terrorism has grown to be a major and persistent issue. According to the most recent Country Reports on Terrorism, explosions and bombings make up 62% of all terrorist incidents (Siberell 2016) and result in the highest fatalities (50%). Terrorists prefer to employ homemade explosives called improvised explosive devices (IEDs), which are comprised of common household objects (Siberell 2016).

IEDs are affordable, simple to make, and discrete. IED components including electrical tapes, electrical wires, batteries, watches, and mobile phones as well as containers like backpacks and gas tanks are examples of post-detonation IED evidence that is frequently found. DNA is transferred to surfaces by contact. Therefore, forensic DNA analysis is an essential technique for locating the criminals who may have put together an IED. However, there is a poor success rate for STR profiling from IEDs (van Oorschot et al. 2010; Hoffmann et al. 2012). Hoffmann et al. (2012) and Foran et al. (2009) both proposed that the heat produced by explosive explosion and deflagration may cause the degradation of the trace DNA present. Numerous substrates, including potential IED components, have been the subject of studies on the deposition, transmission, and persistence of trace DNA as well as on the variables impacting trace DNA (Quinones and Daniel 2012). The likelihood of generating a high-quality STR profile should rise with the amount of DNA retrieved, but there has not been a thorough investigation of the most effective technique for gathering trace DNA. A few swab types were compared in two brief experiments (Brownlow et al. 2012) and two further studies looked into different swab moistening agents (Thomasma and Foran 2013). Only Verdon et al. (2014) evaluated moistening agents and swab types simultaneously utilizing a variety of swabs, bodily fluids, and substrates. Tape lifting has been suggested as a useful method for DNA collection in addition to swabbing (May and Thomson 2009). Although this would happen somewhat slowly (3–40 times slower) for the white blood cells in the buffy coat compared to red blood cells, deionized water could lysed cell by osmotic shock (Parichehreh et al. 2011). The cells might be easier to pick up for the swabs if phosphate-buffered saline (PBS) was used to rehydrate them. Ethanol and isopropanol, on the other hand, might dry up the cells. It is observed that when the moistened swabs were used to rub the dried spots, both alcohols swiftly evaporated from the substrates. Ethanol and isopropanol, on the other hand, dry up the cells (Thomasma and Foran 2013). It should be observed that when the moistened swabs were used to rub the dried spots, both alcohols swiftly evaporated from the substrates. Comparing SDS and Triton-X to other moistening agents, it was discovered that they were able to retrieve more DNA from glass (Phetpeng et al. 2015).

In another paper, the term "Intelligence Forensics DNA" was described as the human cells that are deposited in low copy numbers (LCN) in sections of the evidence that are difficult to interpret and produce important effects which we can define as "Hidden DNA" (Al Snan and Ghayyath 2019). Different hoax improvised explosive devices (IEDs) were used to show effective mode of collecting such as swabbing, tape lifting, wiping and direct cutting of samples.

Prior to gathering evidence in terrorist cases, specialized trained and experienced forensics investigation teams with the following capabilities must be established (Al-Snan 2022):

- experts in ballistics,
- DNA fingerprints,
- fingerprint,
- forensic electrical engineering, and
- forensic photography can all be found at crime scenes.

Each forensic expert plays a critical part in finding the forensic leads that will help determine how IEDs and their components are made, altered, and implanted. To ensure the greatest possible recovery of results, many procedures must be followed before moving further with the forensic collection (Al-Snan 2022):

- Obtain Good Laboratory Practice (GLP) to prevent contamination or cross-contamination of the evidence.
- Put safety first: Ensure that the improvised explosive devices (IEDs) and/or their components (improvised detonators) are modified and made safe by EOD experts.
- Separate the IED's components (such as wires, devices, detonators, etc.), complete the detailed description, and open any sealed or locked devices.
- After opening, take thorough pictures.

By separating the components of IEDs, unsealing electrical devices, and concentrating on modified items, the principle of Intelligence Forensics DNA is established, leading to two main goals: DNA is not only a human identification tool but also explains exactly why DNA is there (i.e., proves or in fact disproves a suspect's role) as shown in Fig. 11.1 (Al Snan and Ghayyath 2019):

In the past 10 years, the use of fired cartridge casings (FCCs) as a DNA source for resolving violent crimes has attracted a lot of attention. An examination of touch DNA on a surface known to obstruct and/or degrade DNA is done via FCC analysis (Elwick et al. 2022). The extraction and

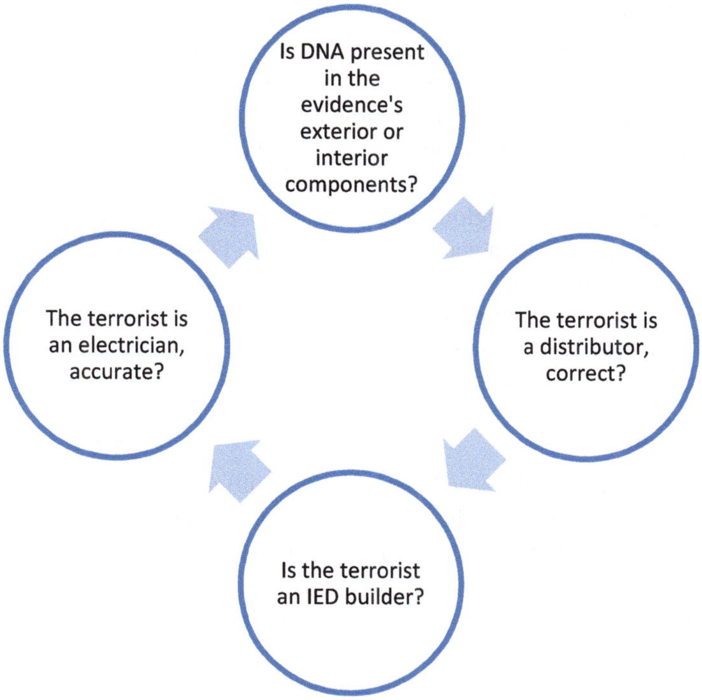

Fig. 11.1 "Touch" DNA can give information leading to the criminal investigation unit better understanding the circumstances for a known terrorist

purification of DNA can lead to losses of more than 60% of the original amount of DNA on the sample, whether DNA analysis is performed immediately or after a delay (Kemp et al. 2014). Direct PCR has thus been suggested as a substitute for touch DNA samples. To increase the amount of template DNA available for STR amplification, direct PCR forgoes the stages of purification and quantification (Martin et al. 2018). However, by omitting purification, active PCR inhibitors might still be present in the sample, which might result in fewer alleles being seen in the electropherograms (Martin et al. 2022). Additionally, quantification is a requirement of the U.S. National DNA Index System's Quality Assurance Standard (QAS) for samples being put into the database (NDIS). It is generally believed that the inhibitory effects of copper ions and degradation from the high temperatures during firing are the main causes of the difficulties in generating DNA profiles from FCCs. Through changes to the majority of the collection, extraction, and purification stages, efforts have been made to improve the recovery of touch DNA. For a more focused sampling, Tonkrongjun et al. employed a fluorescent dye to more accurately pinpoint touch DNA (Tonkrongjun et al. 2019).

Through the use of the twofold swab (wet/dry) approach (Horsman-Hall et al. 2009) and the use of tape lifts (Kanokwongnuwut et al. 2020) in place of swabbing, researchers have modified the sample procedure. In order to improve the likelihood of producing complete profiles, samples have been concentrated using Microcon Centrifugal Filters. Also, since ethylenediaminetetraacetic acid (EDTA) is known to chelate metal ions in solution and decrease the possibility of copper ion coextraction, it has been discovered that adding 0.5 M EDTA to a cotton swab will lessen the impact of copper ions in the extract (Holland et al. 2019).

DNA recovery from such samples are done by two methods: The first technique, called the "soaking" technique after Montpetit and O'Donnell, involves adding a digesting buffer and Proteinase K (ProK) to the cartridge or cartridge case in a 5 mL tube. A single cotton swab is used to remove any residual buffer or DNA after incubation. The lysate and swab are further extracted from and purified (Montpetit and

O'Donnell 2015). The second technique is the "rinse-and-swab" approach, created by Bille et al., which adds bovine serum albumin (BSA) and a tripeptide to the digestion buffer in order to act as a metal ion chelating agent. The buffer is repeatedly washed over the casing before being swabbed with a nylon swab, the same buffer again, and finally a cotton swab leading to DNA co-extracted from both swabs (Bille et al. 2020). For five out of the eight DNA concentrations evaluated on an inert substrate, the soaking method greatly outperformed the rinse-and-swab method in terms of DNA recovery (Elwick et al. 2022).

While several research have looked at different strategies to enhance the recovery and genotyping of low template and degraded DNA samples, few have concentrated particularly on recovering DNA from IED post-blast debris (Berti et al. 2011). Additionally, studies that evaluated DNA recovery from post-blast fragments only paid attention to applying conventional DNA collecting and genotyping techniques (Esslinger 2002). The number of PCR cycles, the injection time during capillary electrophoresis, and the PCR reaction volume have all been recommended as ways to enhance LCN DNA analysis (Budowle and Eisenberg 2009). However, these techniques might amplify stochastic effects, leading to more peaks that stutter, an imbalance in peak height, and alleles that drop in frequency (Børsting et al. 2013). Using alternative molecular markers like insertion/null (INNUL) or single nucleotide polymorphisms (SNPs) are another strategy for extracting more genetic information from difficult DNA samples like those retrieved from IEDs. Long interspersed nuclear elements (LINEs) and short interspersed nuclear elements make up INNULs (SINEs), which are retrotransposable elements (Sinha et al. 2015). INNULs are bi-allelic markers with low mutation rates and no stutter peaks. They may be helpful for amplifying LCN DNA and degraded DNA due to their prevalence in the human genome and capacity to produce tiny amplicons. The commercially available InnoTyper 21 kit (InnoGenomics Technologies) uses a short amplicon (60–125 bp) DNA typing technology and has shown some promise with forensic samples (van den Berge et al. 2016). SNPs are single base alterations that allow for shorter amplicons (150 bp) in the genome, no stutter, and lower mutation rates compared to STRs (Schneider 2012). Similar to INNULs, these features make SNPs suitable for usage with difficult samples such low template and damaged DNA. Furthermore, information can sometimes be extracted from difficult DNA samples using SNPs evaluated using massively parallel sequencing (MPS) (Elena et al. 2016).

11.7 Anti-Narcotics Cases

Every society suffers from the problem of illicit substances, and numerous initiatives have been put in place to address it (Canton 2021). One of the most widely used illegal substances, 3,4-methylenedioxymethamphetamine (MDMA), often known as ecstasy, is typically accessible in tablet form, especially in Europe. Tablets can also contain other substances like methamphetamine, phencyclidine (PCP), and lysergic acid diethylamide (LSD) (Canton 2021). Chemical profiling is currently utilized to analyze illegal narcotics and perhaps link seizures. However, studies have shown that illicit narcotics have a potentially helpful biological component, which may provide law enforcement organizations with new lines of inquiry (Griffin et al. 2021).

For drug paraphernalia like scales, plastic bags, gloves, or trays, DNA profiling is frequently used to identify a person who handled an item (Wong et al. 2019). Previous studies discovered that drug powders may contain human DNA, which could produce a DNA profile (Burgoyne et al. 2012). Moreover, more recent studies have examined the exterior of capsules to find any traces of DNA from anyone who may have handled the capsules during the production or distribution of the medications (Griffin et al. 2021, 2022a). Despite the fact that tablets and capsules may be handled similarly during distribution (drugs are divided into bags for distribution), there may be fewer opportunities for physical handling to occur during the tableting process (which includes a mechanical press) than during capsule preparation. Additionally, compared to tablets (which could be powdery), the naturally sticky gelatine capsule shells could change the amount of DNA deposited after brief manipulation of the substrates.

Tablets differ from capsules in that the drug-containing component is a separate object from the capsule casing, making it easy to separate the component parts for two independent DNA sample procedures. Contrarily, the drug substance is present on the tablet surface, raising the possibility that any DNA that was synthesized and integrated may also be found on the tablet's exterior and vice versa (potentially complicating the profiles). Additionally, tablets expose DNA both on the outside and inside to the drug, which could interfere with subsequent DNA processing or result in DNA destruction. There are several steps in the synthetic production of illegal drugs like MDMA or methamphetamine that may call for the product to be exposed to the outside air. This presents a chance for the product to be exposed to airborne DNA-containing contaminants.

One illustration is when the medication powder is allowed to dry on a tray (typically after either salting out or recrystallisation processes). Depending on the surroundings, including the amount of foot traffic activity, the ventilation, and other factors, the amount of DNA and its origin may be highly changeable (Clare et al. 2021). As even in operational forensic laboratories there are still instances of potential inadvertent contamination occurring despite the use of personal protective equipment and anti-contamination practices, the standards of cleanliness within the covert laboratory and the presence of individuals coughing, sneezing, scratching, and talking near powders may provide opportunities for DNA to be added (Mercer et al. 2022). Trays, glassware, and other frequently handled or exposed things like benches or glassware may provide opportunities for secondary or higher order DNA transfer from contaminated surfaces onto the drug powders. Typically, high-throughput semi- or completely automated processes are used in clandestine laboratories to prepare illicit tablets, and the capacity of these processes depends on the type of tablet press that is employed (i.e., a single or multi-stage press). However, this is probably the stage where the least quantity of DNA is added to the powders.

DNA may be transferred from the objects from which the powders are put into the press and into which the finished tablet is gathered. It is common for large bags of tablets to be seized (Heroin et al. 2021), and at this point, it is expected that most of the DNA would be discovered embedded into the tablet from contamination prior to tableting. The division of the tablets into smaller bags for user distribution may present opportunities for touch DNA from the user's hands to be transferred onto the exterior of the tablet. Many illegal tablets have a logo imprinted on them, which might enable DNA to be concentrated there (Marquis et al. 2008). Even if these people wear gloves, there is still a potential that their DNA would transfer because, according to a study, gloves can contaminate even in forensic laboratories (Goray et al. 2019). It is plausible to assume that the personnel participating in the distribution may not be the same as those involved in the synthetic fabrication if the bigger bags of tablets are delivered to a secondary site before being divided into smaller bags (Griffin et al. 2022b).

Therefore, gathering two distinct DNA fractions—one from the tablet's exterior and another from its interior—might be helpful, especially if information or proof are required to identify the people connected to various stages of the drug production process as shown in Fig. 11.2 (Griffin et al. 2022b).

The research reported here supports the idea that profiling of human biological material obtained from tablets during their creation and subsequent handling may be a significant new advancement in the fight against illicit drugs (Griffin et al. 2022b).

11.8 Sexual Assault Cases

In order to obtain distinct sperm (i.e., male) and non-sperm (e.g., female vaginal epithelial and non-sperm male) fractions of DNA from vaginal

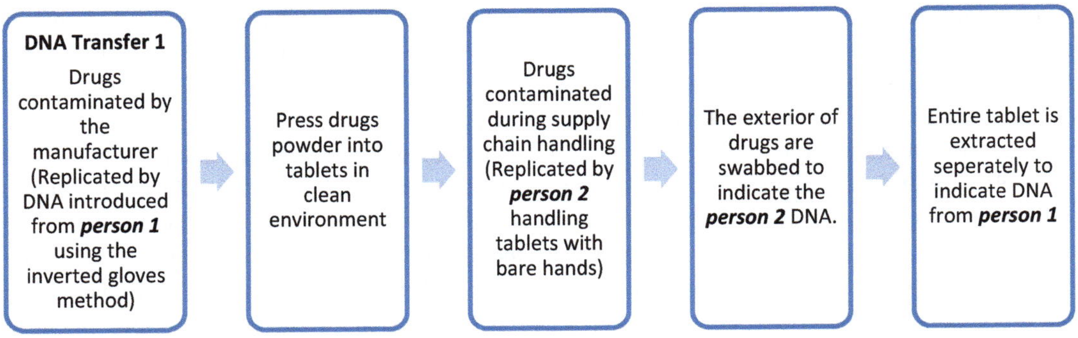

Fig. 11.2 Schematic for experimental design (Griffin et al. 2022b)

swabs taken during the collection of sexual assault evidence, forensic DNA analysis of sexual assault evidence typically requires extraction of the DNA in a differential manner. The victim's DNA can be confirmed using the non-sperm fraction, and the sperm fraction is then used to determine who the offender or sperm donor is. DNA extraction is frequently followed by DNA quantification, PCR amplification of genetic markers (such as STR loci), capillary electrophoresis separation of the PCR products, and data analysis (Sinha et al. 2022).

There is no fully automated, high-throughput technology for the differential DNA extraction method that has been utilized frequently for processing sexual assault material during the past few decades. These difficulties have frequently been named as a significant cause of bottlenecks in forensic laboratories that adds to the significant backlog of unresolved sexual assault cases in countries such as the US (Sinha et al. 2022).

The SpermX method makes use of a technique for physically separating sperm cells from DNA that has been digested by epithelial cells. Numerous bio-separation applications have made use of nanofiber membranes. Fibers with a diameter of less than 1000 nm (1000×10^{-9} m) are referred to as nanofibers (Menkhaus et al. 2010). Interfacial polymerization, electrospinning, or electrostatic spinning can all be used to create nanofibers from a number of materials. Numerous synthetic polymers, including vinyl polymers, acrylic polymers, polyesters, polyether, acrylic polymers, polycarbonate, and polyimides, can be used to create nanofibers. There have been numerous reports of the use of other polymers and copolymers, including various biocompatible polymers like polycaprolactone, in the manufacture of nanofibers (Salem 2001). For the SpermX procedure, the swab is left in the tube after epithelial digestion, thus increasing the sperm recovery when there is very low level of sperm present in the evidence sample.

Additionally, due to a lack of DNA or accidental contamination, forensic samples taken from a crime scene may not always be adequate for DNA typing. As a result, a vital stage in the DNA typing procedure is screening of forensic evidence. The decision to move forward with downstream DNA typing is influenced by the presence or absence of male-specific Y-chromosome DNA (Y-DNA), particularly in sexual assault instances. Quantitative real-time polymerase chain reaction (qPCR) is a popular method for detecting Y-DNA in a variety of forensic samples, including blood, saliva, and touch DNA samples. Microscopic examination of the sperm cells in semen samples and an antigen-based test, such as the prostate-specific antigen (PSA) test, can both identify the male component (Weitzig et al. 2015) independent of qPCR-based Y-DNA detection. Processing many samples quickly is challenging, though, because microscopic analysis requires time and effort. Although the PSA test can be used for high-throughput screening of semen samples, it has issues with false-positive and false-negative results.

As a result, although microscopic examination and the PSA test are required for semen

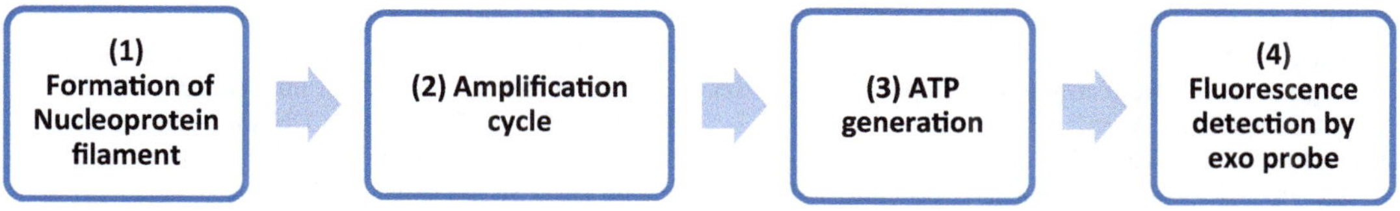

Fig. 11.3 Schematic representation of RPA reaction (Kubo et al. 2022)

identification, they may not be sufficiently quick or precise for screening. Furthermore, for azoospermic semen samples, the screening method based on these two studies is typically unsatisfactory. The absence of a sperm cell does not imply the absence of Y-DNA because Y-DNA, which is produced from epithelial cells and leukocytes, is present even in azoospermic semen (Soares-Vieira et al. 2007). qPCR is essential for template preparation in downstream DNA typing since it is a powerful tool for DNA quantification. However, qPCR analysis is a time-consuming technique that needs DNA extraction, purification, and amplification when utilized for the screening of Y-DNA samples. A quick and easy method for Y-DNA screening is needed to solve these problems. In order to streamline the screening procedure and reduce sample loss, direct Y-DNA detection without the DNA purification step is ideal, as suggested in earlier research (Scott et al. 2019). An isothermal technique for amplifying nucleic acids is recombinase polymerase amplification (RPA). The main benefits of RPA over qPCR are its quick detection time of about 20 min and low operating temperature of 37–42 °C. Recombinase, single-strand DNA binding protein, strand-displacing DNA polymerase, and a few auxiliary proteins combine to form the basis for the RPA process (e.g., T4 UvsY, phosphocreatine, and creatine kinase) (Lobato and O'Sullivan 2018). The process of RPA is shown in Fig. 11.3.

The Y-RPA assay had a detection limit of 0.025–0.005 ng/L for male DNA, and Y-DNA could be found in mixed DNA samples with a 2000-fold excess of female DNA. The current format of the Y-RPA test may not be suited for the screening of low-level DNA samples, nevertheless, because its sensitivity was lower than that of commercially available qPCR kits with detection limits of 1 pg/L (Ewing et al. 2016). Therefore, additional optimization is needed before practical deployment. Three inhibitors' effects on the Y-RPA assay were investigated (humic acid, indigo carmine, and hematin). The Y-RPA assay tolerated inhibitors up to 1000 ng/L humic acid, 250 M indigo carmine, and 500 M hematin, which is comparable to or slightly greater than the qPCR assay (Quantifiler Trio), which is also said to be resistant to inhibitors (Holt et al. 2016). These findings suggest that the Y-RPA assay is a promising tool for the screening of male DNA from forensic samples. As was presented in this chapter, casework is benefitting from ongoing research in the areas of terrorism, sexual assault, drugs, and burglary cases.

References

Al Snan NR, Ghayyath MA (2019) Potential use of touch DNA in terrorism cases: a report of four cases. Arab J Forensic Sci Forensic Med 1(9):1267

Alessandrini F, Cecati M, Pesaresi M, Turchi C, Carle F, Tagliabracci A (2003) Fingerprints as evidence for a genetic profile: morphological study on fingerprints and analysis of exogenous and individual factors affecting DNA typing. J Forensic Sci 48(3):586–592

Al-Snan NR (2022) Potential use of DNA profiling in solving terrorism cases. In: Dash HR, Shrivastava P, Lorente JA (eds) Handbook of DNA profiling. Springer, Singapore, p 589

Barash M, Reshef A, Brauner P (2010) The use of adhesive tape for recovery of DNA from crime scene items. J Forensic Sci 55(4):1058–1064

Berti A, Barni F, Virgili A, Colozza C, Maiorino F, Tocca M (2011) The recovery of DNA profiles from saliva and touch evidences after postal bomb explosion. Forensic Sci Int Genet Suppl Ser 3(1):e471–e472

Bille TW, Fahrig G, Weitz SM, Peiffer GA (2020) An improved process for the collection and DNA analysis of fired cartridge cases. Forensic Sci Int Genet 46: 102238

Børsting C, Mogensen HS, Morling N (2013) Forensic genetic SNP typing of low-template DNA and highly degraded DNA from crime case samples. Forensic Sci Int Genet 7(3):345–352
Brownlow RJ, Dagnall KE, Ames CE (2012) A comparison of DNA collection and retrieval from two swab types (cotton and nylon flocked swab) when processed using three QIAGEN extraction methods. J Forensic Sci 57(3):713–717
Budowle B, Eisenberg AJ, Daal AV (2009) Validity of low copy number typing and applications to forensic science. Croat Med J 50(3):207–217
Burgoyne L, Catcheside D, Kirkbride K, Pearman C (2012) Assessment of the utility of obtaining human profiles from drug seizures. National Drug Law Enforcement Research Fund, Canberra
Canton H (2021) United Nations Office on drugs and crime—UNODC. In: The Europa directory of international organizations 2021. Routledge, London, pp 240–244
Clare E, Economou C, Faulkes C, Gilbert J, Bennett F, Drinkwater R, Littlefair J (2021) eDNAir: proof of concept that animal DNA can be collected from air sampling. PeerJ 9:e11030
Daly DJ, Murphy C, McDermott SD (2012) The transfer of touch DNA from hands to glass, fabric and wood. Forensic Sci Int Genet 6(1):41–46
Dierig L, Schwender M, Wiegand P (2020) Looking for the pinpoint: optimizing identification, recovery and DNA extraction of micro traces in forensic casework. Forensic Sci Int Genet 44:102191
Elena S, Alessandro A, Ignazio C, Sharon W, Luigi R, Andrea B (2016) Revealing the challenges of low template DNA analysis with the prototype Ion AmpliSeq™ Identity panel v2. 3 on the PGM™ Sequencer. Forensic Sci Int Genet 22:25–36
Elwick K, Gauthier Q, Rink S, Cropper E, Kavlick MF (2022) Recovery of DNA from fired and unfired cartridge casings: comparison of two DNA collection methods. Forensic Sci Int Genet 59:102726
Esslinger KJ (2002) Using STR analysis to detect human DNA on exploded pipe bomb devices. Michigan State University, East Lansing
Ewing MM, Thompson JM, McLaren RS, Purpero VM, Thomas KJ, Dobrowski PA, DeGroot GA, Romsos EL, Storts DR (2016) Human DNA quantification and sample quality assessment: developmental validation of the PowerQuant® system. Forensic Sci Int Genet 23:166–177
Farash K, Hanson EK, Ballantyne J (2015) Enhanced genetic analysis of single human bioparticles recovered by simplified micromanipulation from forensic 'touch DNA' evidence. J Vis Exp 97:e52612
Foran DR, Gehring ME, Stallworth SE (2009) The recovery and analysis of mitochondrial DNA from exploded pipe bombs. J Forensic Sci 54(1):90–94
Goray M, Van Oorschot RA, Mitchell JR (2012) DNA transfer within forensic exhibit packaging: potential for DNA loss and relocation. Forensic Sci Int Genet 6(2): 158–166
Goray M, Pirie E, van Oorschot RA (2019) DNA transfer: DNA acquired by gloves during casework examinations. Forensic Sci Int Genet 38:167–174
Gosch A, Courts C (2019) On DNA transfer: the lack and difficulty of systematic research and how to do it better. Forensic Sci Int Genet 40:24–36
Griffin A, Kirkbride KP, Henry J, Painter B, Linacre A (2021) DNA on drugs! A preliminary investigation of DNA deposition during the handling of illicit drug capsules. Forensic Sci Int Genet 54:102559
Griffin A, Kirkbride KP, Henry J, Painter B, Linacre A (2022a) DNA on drugs (part 2): an extended study into the transfer and persistence of DNA onto illicit drug capsules using realistic scenarios. Forensic Sci Int Genet 60:102740
Griffin A, Kirkbride KP, Henry J, Painter B, Linacre A (2022b) Recovery of integrated and surface trace DNA from illicit drug tablets. Forensic Sci Int Genet 61: 102772
Heroin M, C. New and C. Amphetamines (2021) European monitoring centre for drugs and drug addiction (2021) European drug report 2021: trends and developments
Hoffmann SG, Stallworth SE, Foran DR (2012) Investigative studies into the recovery of DNA from improvised explosive device containers. J Forensic Sci 57(3): 602–609
Holland MM, Bonds RM, Holland CA, McElhoe JA (2019) Recovery of mtDNA from unfired metallic ammunition components with an assessment of sequence profile quality and DNA damage through MPS analysis. Forensic Sci Int Genet 39:86–96
Holt A, Wootton SC, Mulero JJ, Brzoska PM, Langit E, Green RL (2016) Developmental validation of the Quantifiler® HP and Trio Kits for human DNA quantification in forensic samples. Forensic Sci Int Genet 21: 145–157
Horsman-Hall KM, Orihuela Y, Karczynski SL, Davis AL, Ban JD, Greenspoon SA (2009) Development of STR profiles from firearms and fired cartridge cases. Forensic Sci Int Genet 3(4):242–250
Kanokwongnuwut P, Kirkbride KP, Linacre A (2020) An assessment of tape-lifts. Forensic Sci Int Genet 47: 102292
Kemp BM, Winters M, Monroe C, Barta JL (2014) How much DNA is lost? Measuring DNA loss of short-tandem-repeat length fragments targeted by the PowerPlex 16® system using the Qiagen MinElute purification kit. Hum Biol 86(4):313–329
Kita T, Yamaguchi H, Yokoyama M, Tanaka T, Tanaka N (2008) Morphological study of fragmented DNA on touched objects. Forensic Sci Int Genet 3(1):32–36
Kubo S, Niimi H, Kitajima I (2022) Rapid and direct detection of male DNA by recombinase polymerase amplification assay. Forensic Sci Int Genet 59:102704
Lehmann V, Mitchell R, Ballantyne K, van Oorschot R (2013) Following the transfer of DNA: how far can it go? Forensic Sci Int Genet Suppl Ser 4(1):e53–e54
Lim S, Subhani Z, Daniel B, Frascione N (2016) Touch DNA—the prospect of DNA profiles from cables. Sci Justice 56(3):210–215

Lobato IM, O'Sullivan CK (2018) Recombinase polymerase amplification: basics, applications and recent advances. TrAC Trends Anal Chem 98:19–35
Locard E (1930) The analysis of dust traces. Am J Police Sci 1:276
Marquis R, Weyermann C, Delaporte C, Esseiva P, Aalberg L, Besacier F, Bozenko JS Jr, Dahlenburg R, Kopper C, Zrcek F (2008) Drug intelligence based on MDMA tablets data: 2. Physical characteristics profiling. Forensic Sci Int 178(1):34–39
Martin B, Blackie R, Taylor D, Linacre A (2018) DNA profiles generated from a range of touched sample types. Forensic Sci Int Genet 36:13–19
Martin B, Plummer A, Linacre A, Henry J (2022) Direct PCR of fired shotgun casings: a South Australian evaluation. Aust J Forensic Sci 54(3):358–364
May R, Thomson J (2009) Optimisation of cellular DNA recovery from tape-lifts. Forensic Sci Int Genet Suppl Ser 2(1):191–192
Menkhaus TJ, Zhang L, Fong H (2010) Applications of electrospun nanofiber membranes for bioseparations. Nova Science Publishers, New York
Mercer C, Henry J, Taylor D, Linacre A (2022) What's on the bag? The DNA composition of evidence bags pre-and post-exhibit examination. Forensic Sci Int Genet 57:102652
Montpetit S, O'Donnell P (2015) An optimized procedure for obtaining DNA from fired and unfired ammunition. Forensic Sci Int Genet 17:70–74
Oleiwi A, Morris M, Schmerer W, Sutton R (2015) The relative DNA-shedding propensity of the palm and finger surfaces. Sci Justice 55(5):329–334
Pang B, Cheung B (2007) Double swab technique for collecting touched evidence. Legal Med 9(4):181–184
Parichehreh V, Estrada R, Kumar SS, Bhavanam KK, Raj V, Raj A, Sethu P (2011) Exploiting osmosis for blood cell sorting. Biomed Microdevices 13:453–462
Petricevic SF, Bright J-A, Cockerton SL (2006) DNA profiling of trace DNA recovered from bedding. Forensic Sci Int 159(1):21–26
Phetpeng S, Kitpipit T, Thanakiatkrai P (2015) Systematic study for DNA recovery and profiling from common IED substrates: from laboratory to casework. Forensic Sci Int Genet 17:53–60
Quinones I, Daniel B (2012) Cell free DNA as a component of forensic evidence recovered from touched surfaces. Forensic Sci Int Genet 6(1):26–30
Salem D (2001) Structure formation in polymeric fibers. Hanser, Munich
Schneider PM (2012) Beyond STRs: the role of diallelic markers in forensic genetics. Transfus Med Hemother 39(3):176–180
Scott A, Jackson K, Carter M, Comeau R, Layne T, Landers J (2019) Rapid sperm lysis and novel screening approach for human male DNA via colorimetric loop-mediated isothermal amplification. Forensic Sci Int Genet 43:102139
Siberell J (2016) Country reports on terrorism 2015. US State Department
Sinha S, Murphy G, Brown H, Montgomery A, Carrol M, Tabak J (2015) Retrotransposable elements: novel and sensitive DNA markers and their application in human identity. Forensic Sci Int Genet Suppl Ser 5:e627–e629
Sinha SK, Brown H, Holt H, Khan M-r, Brown R, Sgueglia JB, Loftus A, Murphy G, Montgomery A (2022) Development and validation of a novel method "SpermX™" for high throughput differential extraction processing of sexual assault kits (SAKs) for DNA analysis. Forensic Sci Int Genet 59:102690
Soares-Vieira JA, Correia Billerbeck AE, Miazato Iwamura ES, Zampieri RA, Fígaro Gattás GJ, Munoz DR, Hallak J, Bilharinho Mendonca B, Lucon AM (2007) Y-STRs in forensic medicine: DNA analysis in semen samples of azoospermic individuals. J Forensic Sci 52(3):664–670
Stanciu CE, Philpott MK, Kwon YJ, Bustamante EE, Ehrhardt CJ (2015) Optical characterization of epidermal cells and their relationship to DNA recovery from touch samples. F1000Res 4:1360
Sun F, Reichenberger EJ (2014) Saliva as a source of genomic DNA for genetic studies: review of current methods and applications. Oral Health Dent Manag 13(2):217–222
Thomasma SM, Foran DR (2013) The influence of swabbing solutions on DNA recovery from touch samples. J Forensic Sci 58(2):465–469
Tonkrongjun P, Phetpeng S, Asawutmangkul W, Sotthibandhu S, Kitpipit T, Thanakiatkrai P (2019) Improved STR profiles from improvised explosive device (IED): fluorescence latent DNA detection and direct PCR. Forensic Sci Int Genet 41:168–176
van den Berge M, Wiskerke D, Gerretsen R, Tabak J, Sijen T (2016) DNA and RNA profiling of excavated human remains with varying postmortem intervals. Int J Legal Med 130:1471–1480
van Oorschot R, Ballantyne K, Mitchell R (2010) Forensic trace DNA: a review. Investig Genet 1(1):1–17
Verdon TJ, Mitchell RJ, van Oorschot RA (2014) Swabs as DNA collection devices for sampling different biological materials from different substrates. J Forensic Sci 59(4):1080–1089
Weitzig L, Schroeder AS, Augustin C, Raupach T, Sehner S, Anders S (2015) Diagnostic value of PSA and AP tests for the detection of spermatozoa in postmortem swabs from the genital and anal region in males. J Forensic Sci 60(1):41–44
Wickenheiser RA (2002) Trace DNA: a review, discussion of theory, and application of the transfer of trace quantities of DNA through skin contact. J Forensic Sci 47(3):442–450
Wong HY, Tan J, Lim ZG, Kwok R, Lim W, Syn CK-C (2019) DNA profiling success rates of commonly submitted crime scene items. Forensic Sci Int Genet Suppl Ser 7(1):597–599
Zoppis S, Muciaccia B, D'Alessio A, Ziparo E, Vecchiotti C, Filippini A (2014) DNA fingerprinting secondary transfer from different skin areas: morphological and genetic studies. Forensic Sci Int Genet 11: 137–143

12 Future Directions of Forensic DNA Analysis

12.1 Introduction

Since its inception in the mid-1980s, forensic DNA analysis techniques have progressed much in last few decades. The first decade (1985–1995) is considered as the exploration phase where various techniques such as RFLP, PCR, STR markers were standardized and the quality control parameters were assessed. The second decade of forensic DNA analysis, during 1995–2005 saw stabilization and standardization of the techniques. In this decade, various national DNA databases were established, several multiplex systems were discovered and capillary electrophoresis (CE) based analysis were developed. In the third decade during 2005–2015, the technique showed growth and advanced in form of rapid growth of DNA databases, rise in the use of Y-STRs, and advanced research works in the field of familiar searching and NGS analysis. The last decade has showed tremendous advancements in the field of highly sensitive techniques, probabilistic approaches to result interpretation, allele sequencing, and much more (Butler 2015) and continues in this way as this book has shown.

Forensic DNA analysis has wider applications than expected. Any forensic analysis can be used for either investigative, predictive, and preventive purpose (Fig. 12.1). Since its inception, forensic DNA analysis has been used for investigative purposes. The technique has proved its significance in investigating various criminal cases where biological samples are involved. After the development of NGS techniques, the predictive approach of forensic DNA analysis has been explored extensively. It enables a DNA analyst to predict the biogeographic ancestry, phenotypic features, and lineage of an individual (Schneider et al. 2019). It helps in providing useful investigative leads to the investigating agencies by predictive approach. However, in comparison to the investigative and prediction approach, the preventive approach of forensic DNA analysis is explored only a little. Forensic intelligence has been used in analysis of data obtained on various criminological parameters and such data have been translated into actionable intelligence (Delgado et al. 2021). In a similar fashion, forensic DNA analysis is being explored for prevention of crime, mostly in the realm of sexual assault cases (Wickenheiser 2021).

With the advancement of forensic DNA analysis, the technique is also facing huge problems. Individual privacy concern is one of the major long debated issues in forensic DNA analysis. Nonconsensual DNA sampling and retention of DNA profiles in databases has been considered as a violation to human rights and raising privacy concern (Williams and Johnson 2005). With the increased sensitivity of techniques, low copy number (LCN) samples are beingly readily amplified and analyzed. LCN analysis results are not reproducible and seem inconsistent (Budowle et al. 2009). The chance of accidental transfer of DNA can also lead to incorrect prosecution of an

H. R. Dash et al., *Advancements in Forensic DNA Analysis*,
https://doi.org/10.1007/978-981-99-6195-5_12

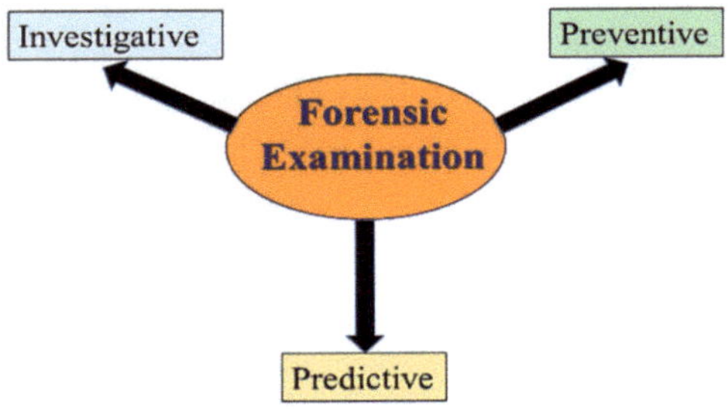

Fig. 12.1 Significant approaches of forensic DNA analysis

innocent individual. Attribution of a DNA profile to an individual has been well established and readily accepted by the courts. However, how the DNA profile landed in the object is being debated in the courts now-a-days (Roux et al. 2021). Keeping this in view, futuristic applications of forensic DNA analysis are the topic of discussion in this chapter.

12.2 Faster Results

In last few years, RAPID DNA analysis has gained popularity in generating a DNA profile in less than 2 h. Integrated systems have been developed to carry out the routine forensic DNA workflow such as DNA extraction, amplification of STR markers, capillary electrophoresis, and analysis of separated DNA fragments. Such platforms are completely automated and can be transported to the field to carry out the required tests (Ragazzo et al. 2020). Though such instruments are capable of developing a DNA profile in less than 2 h, the forensic DNA fraternity is more optimist to further reduce the time requirement for such tests. Hopefully, in the future, generation of a DNA profile is possible as instantly as a web search.

Though these advanced techniques are capable of generating a DNA profile in much less time, the cost involvement in these tests is very high. Currently, the reagent expenses of such rapid DNA tests are about ten times higher than that of the conventional techniques (Butler 2015). Thus, in the future DNA analysts speculate about having a sophisticated instrument capable of generating a DNA profile instantly with high cost-effectiveness.

12.3 High Sensitivity and High Information

In order to generate more information from STR DNA profiles, the original 13 CODIS loci have been expanded to 20 CODIS. Increasing in the number of STR markers reduces the random match probability and thus increases the effectiveness of the individualization capability of a DNA profile. But what after 10–15 years? Should we keep on increasing the number of STR markers? Hence, for some, the time has come to include population specific quality markers. In this regard, a study by Dash et al. (2020) showed that the forensic and paternity parameters of a set of 19 STR markers showed greater usefulness in comparison to the 20 recommended expanded-CODIS markers in the central Indian population. This set of 19 STR markers included three non-CODIS markers, i.e., Penta D, Penta E and SE33. In the future, we may expand to population-specific useful STR marker sets rather than keeping on increasing the number of STR markers year by year.

The limitation of the Capillary Electrophoresis technique is also a major constraint in accommodating a greater number of STR markers. The dye-based chemistry allows simultaneous detection of few STR markers simultaneously. Most of the forensic DNA analysis laboratories use 6-dye chemistry for simultaneous detection of multiple STR markers. However, recently Promega Corp. developed an eight-dye instrument which capable of separating and distinguishing a greater number of STR markers simultaneously (Burgardt and Weissenberger 2021). When an increased number of STR markers are needed to be analyzed, present-day capillary electrophoresis instruments may not solve the purpose. However, next generation sequencing (NGS) technology is capable of analyzing a greater number of STR markers simultaneously. With the advent of NGS technology, SNP markers have started playing a pivotal role in generating more useful information mostly in terms of investigative leads. NGS technology allows the simultaneous analysis of a large set of SNP markers which facilitates the accurate prediction of much useful information such as ancestry, lineage, and phenotype (Schneider et al. 2019).

Another major problem faced by the forensic DNA analysts is sex determination in a DNA profile. Amelogenin if the established sex determining marker for forensic use. The size difference between AMELX (106 bp) and AMELY (112 bp) allows the determination of sex in a DNA profile. However, many populations show the phenomenon of AMELY deletion, especially the Indian, Japanese, and Italian populations (Turrina et al. 2009). Determination of the actual sex of an individual using the conventional amelogenin marker is immensely difficult. To overcome such problems, the commercially available STR kits have started using some additional markers such as DYS391, DYS570, DYS576, and/or YInDel for determination of sex in AMELY deletion cases. However, when AMELY deletion occurs, DYS570 and DYS576 markers also do not amplify as they are present in the AMELY region on the Y chromosome (Kareem et al. 2015). Besides, the DYS391 marker is present in close proximity to the AMELY region and is more prone to deletion. Microdeletions in YInDel marker have also been reported. Besides, these additional markers are amplified only in male samples. Thus, we may require an alternative sex determining marker which gets amplified in both male and female samples as well as differentiating between them. In this regard, Dash et al. (2020) proposed few alternative markers such as STS, SRY, TSPY, DXYS156, SNPs, DYZ1 for accurate sex determination which is also useful in AMELY deletion cases. In the future, many such alternative sex determining markers are expected to be developed and available commercially.

12.4 Analysis of Challenging Samples

The present-day forensic DNA analysis is facing a huge problem with the analysis of challenging low template samples and the biological samples containing a huge quantity of PCR inhibitors. Recent advancements in the form of shortening the amplicon size by the use of mini-STRs and using small size markers such as SNPs and InDels seek to ameliorate these issues (Cortellini et al. 2019). To meet to sufficient level of discrimination with that obtained with the STR markers, a larger number of such SNP and InDel markers can be analyzed in NGS platforms.

Analysis of complex DNA mixtures is another important challenge for a forensic DNA examiner. Though recent advancements have shown the development of probabilistic genotyping software for the analysis of complex mixture cases, they are of limited use (Holland et al. 2022). With the increase in the number of contributors, the mixture interpretation becomes highly challenging. The situation becomes worse when any of the contributors is biologically related. The allele sharing among the biological relatives further complicates their attribution in a mixed DNA profile. Many DNA examiners use the probabilistic genotype software as "black box" to report the finding without even understanding it properly. Thus, in the future, a reliable DNA report is deemed necessary containing the limitations of methods, models, assumptions made and interpretation applied with the case results (Butler 2015).

12.5 Investigative Genetic Genealogy

In the last one-decade investigative genetic genealogy (IGG) has made a remarkable advancement in the field of forensic DNA analysis. As discussed in the previous chapters, the conventional forensic DNA analysis matches the DNA profile between the questioned and reference sample for finding a possible match. However, at many instances, reference samples are not found. This condition arises due to many reasons including no suspect, no eyewitness or cold cases. In this regard, familial searching can be carried out to find the possible match among pairwise samples for the first cousins (12.5% similarity), second cousins (3.13% similarity), third cousins

(0.78% similarity), and beyond. Such familial searching techniques are became available in the last 10 years by analyzing more than 600,000 SNPs using high-density microarray genotyping techniques. In contrast to familial searching, genealogy allows scientists to establish a relationship between profiles by analyzing the genetic distance among the individuals (Perego et al. 2019).

IGG describes the use of SNP based matching combined with family tree search to generate investigative leads in criminal investigations and cases of missing person identification. To achieve a high degree of effectiveness using the IGG technique, genotype data on a large number of SNP markers should be available at very low cost. Other requirements are the availability of a large database of SNP genotypes with public access and a simple well-established technique to compare and analyze the genotype results of the large-scale SNP data (Kling et al. 2021). Different private service providers have explored the use of IGG such as 23andMe, Family Finder test, AncestryDNA, and MyHeritageDNA. The features of various IGG service providers is given in Table 12.1. The use of IGG has shown its importance by solving a huge number of cold cases recently. However, as a global perspective when many countries do not have a proper STR DNA database, the genome-wide genotype database of SNP markers is a futuristic dream. Looking into the significance and utility of the IGG approach, we might witness a significant increase in the number of cases solved by IGG and expansion of this technique around the globe with high numbers of the population using direct-to-consumer genetic testing kits.

12.6 Microbial Forensics

Microorganisms are omnipresent, omnipotent, and omniscient. However, their forensic use has been limited to this date. Since ancient times, pathogenic microorganisms have been used for bioweapon purposes. Recently, microbial forensic research has shown expansion into body-fluid identification, prediction of geolocation of an individual, investigation in sexual assault cases, prediction of post-mortem time interval, and many more applications.

Detection of body fluids using commensal signature microorganisms has been explored recently. Several indicator microorganisms have been detected in various body fluids. These include *Veillonella atypica*, *Streptococcus mutans* and *Streptococcus salivarius* in saliva, and *Lactobacillus crispatus,* and *Lactobacillus gasseri* in vaginal fluid. Microbial group specific primers have also been developed to identify the targeted microorganisms to detect the body fluids. Similarly, microorganisms can play a major role in determining the cause of death. Though diatoms are established biomarkers for distinguishing between postmortem and antemortem drowning, such tests are associated with several limitations. Some diatoms with large size cannot penetrate the alveolar membrane and cannot be detected in the soft and hard tissues whereas, microorganisms can pass through the alveolar membrane to show their presence in various tissues in antemortem drowning cases. The indicator microorganisms in drowning case investigation include *Synechococcus*, *Melosira*, *Staurastrum*, *Synechocystis*, *Chlorogloeopsis*, *Anabaena*, *Chroococcidiopsis*.

Certain microorganisms are responsible for the generation of several metabolites as a by-product during the conversion of several toxicological substances in the body. For example, *Bacillus cereus*, *Staphylococcus epidermidis*, *Clostridium perfringens,* and *Bacteroides fragilis* are known to be responsible for bioconversion of nitrobenzodiazepines. Similarly, gastrointestinal microorganisms Enterobacterium, Enterococcus, Lactobacillus, Clostridium, and Bacteroides are responsible for the N-demethylation of methamphetamines by converting these compounds into amphetamines. Detection of these by-products and the associated microorganisms are also useful in various forensic toxicological analysis. Similarly, death due to hospital-acquired or nosocomial infections can also be established by the detection of common pathogens including *Acinetobacter baumannii,*

Table 12.1 The commercially available service providers of Investigative Genetic Genealogy (adopted from Kling et al. 2021)

	23andMe	Ancestry.com	FTDNA	GEDmatch	MyHeritage
Website	www.23andme.com	www.ancestry.com/dna	www.familytreedna.com	www.gedmatch.com	www.myheritage.com/dna
Establishment	2006	1996	2000	2010	2003
International availability	50+ countries	34 countries	All countries except Iran and Sudan	Worldwide	All countries except Israel, Iran, Libya, Sudan, Somalia, North Korea, Lebanon and Syria
Database Size	12 million	19 million	1.4 million	1.45 million	4.5 million
Chip used	Illumina GSA	Illumina OmniExpress	Illumina GSA	n/a	Illumina GSA
Total SNPs	654,027	700,000	654,027	n/a	654,027
Autosomal SNPs	621,575	637,639	621,575	n/a	621,575
X-SNPs	27,176	28,892	27,176	n/a	27,176

Pseudomonas aeruginosa, Staphlococcus aureus, Klebsiella pneumoniae, *Enterococcus faecium*, *Salmonella enterica,* and *Clostridium difficile*. Besides, Sudden Infant Death (SIDS) and Shaken baby Syndrome is often associated with human herpes virus, Epstein–Barr virus, cytomegalovirus, respiratory syncytial virus, adenovirus, human metapneumovirus, *Neisseria meningitidis*, *Haemophilus influenzae*, *Streptococcus pneumoniae,* and *Bordetella pertussis*. Thus, the analysis of these microorganisms is useful in establishing the cause of death in SIDS.

Microbial mediated determination of time since death is being explored greatly. Both intrinsic and exogenous microorganisms are responsible for the degradation of a human corpse. The human cadaver serves as a great source of nutrients for the microorganisms to invade and grow on it. As the immune system of the body stops functioning soon after the death, microorganisms colonize and facilitate the degradation of a human cadaver. Thanatomicrobial analysis describes the microbial community dynamics in the internal human organs, whereas, necrobiome categorically describes the external microbial communities associated with a human cadaver (Dash and Das 2020). Studies have found that *Proteobacteria*, *Pseudomonas*, *Enterobacteriaceae* are the signature microflora found during the fresh (24–72 h) stage of a human cadaver. Similarly, *Proteobacteria*, *Wohlfahrtiimonas*, *Ignatzschineria* are more prevalent in the bloat stage (3–5 days), *Proteobacteria*, *Pseudomonadaceae*, *Chromatiaceae*, *Proteus* are more prevalent in the active decay stage (8–10 days), *Firmicutes*, *Sporosarcina*, *Lactobacillales*, *Acinetobacter*, *Planococcaceae* are prevalent in the advanced decay stage (few weeks), whereas, *Acidobacteria*, *Bacteriodetes*, *Firmicutes*, and Soil microflora are more prevalent in dry remains (after 1 month) (Lee Goff 2009).

DNA analysis of protective sexual assault cases such as use of condoms and femidoms is highly difficult. Without the transfer of body fluids, results cannot be generated by forensic DNA analysis. A recent study showed that the transfer of female microbiome during intercourse is more significant. Such microbial transfer between heterosexual couples after sexual intercourse allows the detection of identification of the perpetrator in a sexual assault case (Dixon et al. 2023). Several studies have also shown the presence of unique commensal microflora associated with an individual. This speculates that the individualization can also done by analyzing the microbial profile of an individual. Useful research work carried out in recent years allows the exploitation of forensic microbial analysis to move beyond the bioweapons and to provide a great insight in forensic cases.

12.7 CRISPR-Cas9 Technique and Its Challenge

Though Clustered Regularly Interspaced Short Palindromic Repeats (CRISPR) sequences have been discovered in many species of bacteria including *E. coli*, *Salmonella typhi*, *Mycobacterium tuberculosis*, and *Haloferax mediterranei* four decades ago, their exact function was discovered recently. CRISPR is characterized by the 21–37 bp long DNA sequences interspaced by similar-sized non-repetitive DNA sequences. These sequences rely on the Cas9 enzyme with a nuclease activity which is coded by the short CRISPR RNAs (crRNAs) transcribed from the spacer sequences (Pickar-Oliver and Gersbach 2019). Most of the CRISPR-Cas9 associated programming associated with the genome editing is explored for therapeutic applications, however, they may have a potential use in forensic DNA analysis as well.

LCN DNA analysis is a huge problem in forensic DNA analysis. CRISPR-Cas9 technology dependent STR-Seq method has been discovered for generating microsatellite-spanning sequence reads with high coverage and accurate genotypes. In this technique, CRISPR–Cas9 guide RNAs (gRNAs) are synthesized to selectively cut the genomic DNA at a position flanking a targeted STR locus. A single adaptor library is prepared and 40-mer primer probes are used to mediate STR targeting and finally, the sequencing is carried out. Thus, the CRISPR-Cas9 system can be of superior use in forensic DNA analysis, mostly in Low Copy Number (LCN) samples by cutting the DNA fragment at a specific location and carrying out a PCR independent sequencing of the target region. Similarly, this technique has a wider application in analysis of degraded DNA and interpretation of mixed profiles (Dash and Arora 2022).

Despite several advantages of the CRISPR-Cas9 technique and its wider application in forensic DNA analysis, the challenges are also high. From a forensic DNA analysis point of view, CRISPR-Cas technology poses a huge threat in context of modification of a suspect's DNA to evade the process of DNA-based investigation. Owing to such challenges, the forensic DNA analysis techniques should be well equipped to deal with such problems that may arise in the future.

12.8 Triparental Child, Transplantations, and IVF

With the rapid use of in vitro fertilization and sperm and ovum donors, the future of forensic DNA analysis is highly challenging. The concept of the biological father and legal father is changing at a rapid pace. Acceptance of donor eggs and donor sperm is also increasing among infertile couples (Banerjee and Singla 2018). In the future, the analysis of cases such as disaster victim identification is expected to be highly challenging. Without knowing the putative biological father and mother of an individual, it is not possible to conduct a DNA test for possible identification. Similarly, organ transplantation cases may show the presence of two types of DNA in the body of an individual. Though the foreign DNA quantity is relatively low in comparison to the host's DNA profile, they exist for years inside the host's body and are tissue-specific. Such phenomenon is called as microchimerism. Studies have shown the presence of microchimerism in host's body in case of kidney or liver transplant cases for 2.5–30 years postoperatively (Starzl et al. 1997). As most of the multiplex STR kits used now-a-days have a high degree of sensitivity, such microchimerisms are readily detected and may complicate the analysis of DNA profiles.

Besides autosomal and gonosomal STR analysis, mitochondrial analysis is also facing a huge problem due to the recently developed triparental child concept. The triparental child concept is

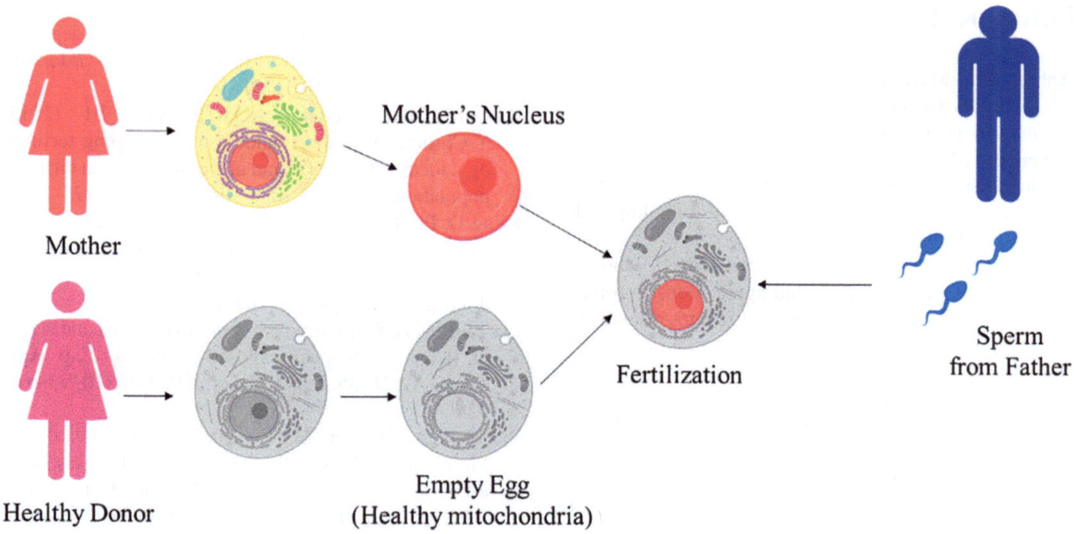

Fig. 12.2 Overview of the formation of a triparental child

being practiced as a mitochondrial replacement therapy. The 16,569 bp circular mitochondrial genome is maternally inherited and contains 37 genes: 13 messenger RNAs, 22 RNAs, and two ribosomal RNAs. Mitochondrial diseases can affect organs that depend on energy metabolism, such as skeletal muscle, cardiac, central nervous system, endocrine, retina and liver, giving rise to several incurable diseases, such as deafness, diabetes mellitus, myopathies, glaucoma and others (Costa et al. 2016). Mitochondrial DNA is exclusively inherited from the female gametes. Thus, the genetic mutations found on a maternal mitochondrial DNA chromosome can be avoided by conducting mitochondrial replacement therapy. The procedure of triparental baby formation in vitro is described in Fig. 12.2. The mother's nucleus is extracted from the egg. The empty egg is extracted from a healthy donor. Fertilization takes place with the biological mother's nucleus, donor's anucleated egg, and the biological father's sperm (Farnezi et al. 2020). In this way, three individuals contribute to the genetic material of the child, i.e., genomic DNA from both the biological mother and father, and mitochondrial DNA from the healthy egg donor. With the rise of such cases, the analysis of mitochondrial DNA to establish maternal lineage of an individual will be complicated.

12.9 Preventive Forensics

Preventive forensic s is regarded as the future of forensic applications of criminal cases. Based on the past experiences and investigative leads provided by the forensic examiners, many crimes can be prevented before it actually happens in a real scenario (Wickenheiser 2021). Though forensic DNA analysis in the field of preventive forensics is in its nascent stage, it is deemed to be a superior approach in the future. Aggression is a common phenomenon in committing a crime. Abnormal aggression is a hallmark of neuropsychiatric disorders and can be elicited by environmental factors acting on an underlying genetic susceptibility. Characterization of such aggression characteristics may provide vital information in preventing crime that could be committed by such individuals. Many genes associated with aggression have been identified including the tryptophan hydroxylase, serotonin 5HT-2A and 5HT-2C receptors, and the serotonin transporter genes (Anholt and Mackay 2012). More research on such population specific genetic association of aggression could be useful in preventing crime. Educational materials will need to continue to be updated as advancements in forensic science continue.

References

Anholt RRH, Mackay TFC (2012) Genetics of aggression. Annu Rev Genet 46:145–164. https://doi.org/10.1146/annurev-genet-110711-155514

Banerjee K, Singla B (2018) Acceptance of donor eggs, donor sperms, or donor embryos in Indian infertile couples. J Hum Reprod Sci 11:169–171. https://doi.org/10.4103/jhrs.JHRS_63_17

Budowle B, Eisenberg AJ, van Daal A (2009) Validity of low copy number typing and applications to forensic science. Croat Med J 50:207–217. https://doi.org/10.3325/cmj.2009.50.207

Burgardt N, Weissenberger M (2021) First experiences with the spectrum compact CE system. Int J Legal Med 136:55–72. https://doi.org/10.1007/s00414-021-02673-1

Butler JM (2015) The future of forensic DNA analysis. Philos Trans R Soc Lond B Biol Sci 370:20140252. https://doi.org/10.1098/rstb.2014.0252

Cortellini V, Correa HSD, Brescia G, Cerri N, Verzeletti A (2019) 38 Insertion-deletion (InDel) markers analysis and its application in forensic cases. Forensic Sci Int Genet Suppl Ser 7:791–793. https://doi.org/10.1016/j.fsigss.2019.10.179

Costa CM, Santana FM, Vilanova SM, Parente DM (2016) Analysis of diseases related to mitochondrial DNA: a literature review. Rev Saúde Foco 3:107–123

Dash HR, Arora M (2022) CRISPR-CasB technology in forensic DNA analysis: challenges and solutions. Appl Microbiol Biotechnol 106:4367–4374. https://doi.org/10.1007/s00253-022-12016-8

Dash HR, Das S (2020) Thanatomicrobiome and epinecrotic community signatures for estimation of post-mortem time interval in human cadaver. Appl Microbiol Biotechnol 104:9496–9512. https://doi.org/10.1007/s00253-020-10922-3

Dash HR, Rawat N, Das S (2020) Alternatives to amelogenin markers for sex determination in humans and their forensic relevance. Mol Biol Rep 47:2347–2360. https://doi.org/10.1007/s11033-020-05268-y

Delgado Y, Price BS, Speaker PJ, Stoiloff SL (2021) Forensic intelligence: data analytics as the bridge between forensic science and investigation. Forensic Sci Int Synerg 3:100162. https://doi.org/10.1016/j.fsisyn.2021.100162

Dixon R, Egan S, Hughes S, Chapman B (2023) The Sexome—a proof of concept study into microbial transfer between heterosexual couples after sexual intercourse. Forensic Sci Int 348:111711. https://doi.org/10.1016/j.forsciint.2023.111711

Farnezi HCM, Goulart ACX, dos Santos A, Ramos MG, Penna MLF (2020) Three-parent babies: mitochondrial replacement therapies. JBRA Assist Reprod 24:189–196. https://doi.org/10.5935/1518-0557.20190086

Holland MM, Tiedge TM, Bender AJ, Gaston-Sanchez SA, McElhoe JA (2022) MaSTR™: an effective probabilistic genotyping tool for interpretation of STR mixtures associated with differentially degraded DNA. Int J Legal Med 136:433–446. https://doi.org/10.1007/s00414-021-02771-0

Kareem MA, Hussein AO, Hameed IH (2015) Y-Chromosome short tandem repeat, typing technology, locus information and allele frequency in different population: a review. Afr J Biotechnol 14:2175–2178

Kling D, Phillips C, Kennett D, Tillmar A (2021) Investigative genetic genealogy: current methods, knowledge and practice. Forensic Sci Int Genet 52:102474. https://doi.org/10.1016/j.fsigen.2021.102474

Lee Goff M (2009) Early post-mortem changes and stages of decomposition in exposed cadavers. Exp Appl Acarol 49:21–36. https://doi.org/10.1007/s10493-009-9284-9

Perego UA, Bodner M, Raveane A, Woodward SR, Montinaro F, Parson W, Achilli A (2019) Resolving a 150-year-old paternity case in Mormon history using DTC autosomal DNA testing of distant relatives. Forensic Sci Int Genet 42:1–7. https://doi.org/10.1016/j.fsigen.2019.05.007

Pickar-Oliver A, Gersbach CA (2019) The next generation of CRISPR–Cas technologies and applications. Nat Rev Mol Cell Biol 20:490–507

Ragazzo M, Melchiorri S, Manzo L, Errichiello V, Puleri G, Nicastro F, Giardina E (2020) Comparative analysis of ANDE 6C rapid DNA analysis system and traditional methods. Genes (Basel) 11:582. https://doi.org/10.3390/genes11050582

Roux C, Willis S, Weyermann C (2021) Shifting forensic science focus from means to purpose: a path forward for the discipline? Sci Justice 61:678–686. https://doi.org/10.1016/j.scijus.2021.08.005

Schneider PM, Prainsack B, Kayser M (2019) The use of forensic DNA phenotyping in predicting appearance and biogeographic ancestry. Dtsch Arztebl Int 116:873–880. https://doi.org/10.3238/arztebl.2019.0873

Starzl TE, Demetris AJ, Murase N, Trucco M, Thomson AW, Rao AS, Fung JJ (1997) Chimerism after organ transplantation. Curr Opin Nephrol Hypertens 11:169–171. https://doi.org/10.4103/jhrs.JHRS_63_17

Turrina S, Filippini G, De Leo D (2009) Evaluation of deleted region from Yp11.2 of two amelogenin negative related males. Forensic Sci Int Genet Suppl Ser 2:240–241. https://doi.org/10.1016/j.fsigss.2009.08.157

Wickenheiser RA (2021) The value of forensic DNA leads in preventing crime and eliminating the innocent. Forensic Sci Int Synerg 3:100201. https://doi.org/10.1016/j.fsisyn.2021.100201

Williams R, Johnson P (2005) Inclusiveness, effectiveness and intrusiveness: issues in the developing uses of DNA profiling in support of criminal investigations. J Law Med Ethics 33:545–558. https://doi.org/10.1111/j.1748-720x.2005.tb00517.x

MIX
Papier aus verantwortungsvollen Quellen
Paper from responsible sources
FSC® C105338

If you have any concerns about our products,
you can contact us on
ProductSafety@springernature.com

In case Publisher is established outside the EU,
the EU authorized representative is:
Springer Nature Customer Service Center GmbH
Europaplatz 3, 69115 Heidelberg, Germany

Printed by Libri Plureos GmbH
in Hamburg, Germany